Experimental and model-based investigation of overpotentials during oxygen reduction reaction in silver-based gas-diffusion electrodes

Experimental and model-based investigation of overpotentials during oxygen reduction reaction in silver-based gas-diffusion electrodes

Doctoral Thesis
(Cumulative Dissertation)

to be awarded the degree
Doctor of Engineering (Dr.-Ing.)

submitted by
David Franzen
from Münster

approved by the
Faculty of Mathematics/Computer Science and Mechanical Engineering
Clausthal University of Technology

date of oral examination
24.09.2021

Bibliografische Information der Deutschen Nationalbibliothek
Die Deutsche Nationalbibliothek verzeichnet diese Publikation in der Deutschen Nationalbibliografie; detaillierte bibliographische Daten sind im Internet über http://dnb.d-nb.de abrufbar.
1. Aufl. - Göttingen: Cuvillier, 2021
Zugl.: (TU) Clausthal, Univ., Diss., 2021, D104

Dean
Prof. Dr. rer. nat. Jörg P. Müller

Chairperson of the Board of Examiners
Prof. Dr.-Ing. Volker Wesling

Supervising tutor
Prof. Dr.-Ing. Thomas Turek

Reviewer
Prof. Dr.-Ing. Ulrike Krewer
apl. Prof. Dr.-Ing. Ulrich Kunz

© CUVILLIER VERLAG, Göttingen 2021
Nonnenstieg 8, 37075 Göttingen
Telefon: 0551-54724-0
Telefax: 0551-54724-21
www.cuvillier.de

Alle Rechte vorbehalten. Ohne ausdrückliche Genehmigung des Verlages ist es nicht gestattet, das Buch oder Teile daraus auf fotomechanischem Weg (Fotokopie, Mikrokopie) zu vervielfältigen.
1. Auflage, 2021
Gedruckt auf umweltfreundlichem, säurefreiem Papier aus nachhaltiger Forstwirtschaft.

ISBN 978-3-7369-7514-9
eISBN 978-3-7369-6514-0

Abstract

Chlorine is one of the most important basic chemicals, which is used directly or indirectly in the production of around 60 % of all chemical products. In 2017, the annual production volume was around 89 million metric tons. The production is almost exclusively based on energy intensive electrolysis processes, with an average of 2.5 - 3.5 MWh of electrical energy required per ton of chlorine generated. This means that chlorine production alone accounts for around 3 % of the electrical energy used worldwide in industry. By using oxygen depolarized cathodes (ODC), it is possible to reduce the demand for electrical energy on an industrial scale by about 25 %. To achieve this, the reaction at the cathode of the electrolysis cell is exchanged. Instead of hydrogen evolution, oxygen reduction takes place. This results in a reduction of the cell voltage by approx. 1 V. Under the given electrolysis conditions (80 °C), however, oxygen has a low solubility in the electrolyte (30 – 32 wt.% NaOH). Therefore, it is necessary to design the ODC as a gas-diffusion electrode (GDE). These are porous silver-based electrodes with hydrophobic regions due to the use of polytetrafluoroethylene (PTFE). During operation, the liquid electrolyte penetrates the pore structure, but the PTFE prevents complete flooding of the electrode. Oxygen is supplied via a gas compartment and enters the internal structure of the electrode. A three-phase interface is formed, consisting of liquid electrolyte, gas and catalytically active solid, at which the electrochemical reaction takes place. Although the technology is already successfully used industrially, many processes, especially the electrolyte distribution, inside the GDE remain unknown.

To gain more insight into the functioning of the GDE, the fraction of PTFE was systematically varied during fabrication and its influence on both the pore system and the electrochemical performance was investigated by means of half-cell experiments. This allows assumptions about the degree of flooding and thus on the electrolyte distribution. However, this also provides only limited insights. Via radiographs using synchrotron radiation during operation in the half-cell, an operando electrolyte distribution could be determined for the first time in a silver-based GDE. By implementing this in an improved so-called thin-film flooded agglomerate (TFFA) model, the processes within the electrode could be analyzed. This provides detailed insights into the processes, e.g. a complex water cycle was identified by which hydroxide ions are removed from the GDE, thus improving the performance of the electrode. However, the penetration depth of the electrolyte cannot be determined by operando experiments so far. For this purpose, graded electrodes were fabricated in further experiments. By using inactive diffusion layers, it was possible to narrow down the reactive zones of the GDE and thus determine the effective penetration depth of the electrolyte. This showed that much larger regions of the electrode must be active than previously assumed.

Kurzfassung

Chlor ist eine der wichtigsten Grundchemikalien und wird bei der Produktion ca. 60 % aller chemischen Produkte direkt oder indirekt eingesetzt. Im Jahr 2017 lag die jährliche Produktionsmenge bei ca. 89 Millionen Tonnen. Die Herstellung erfolgt fast ausschließlich durch energieintensive Elektrolyseverfahren, wobei pro produzierter Tonne Chlor durchschnittlich 2,5 – 3,5 MWh an elektrischer Energie benötigt wird. Damit entfallen alleine auf die Chlorproduktion etwa 3 % der weltweit industriell genutzten elektrischen Energie. Durch den Einsatz von Sauerstoffverzehrkathoden (SVK) ist es möglich, den Bedarf an elektrischer Energie im industriellen Maßstab um ca. 25 % zu senken. Hierfür wird die Reaktion an der Kathode der Elektrolysezelle ausgetauscht. Anstatt der Wasserstoffentwicklung findet die Sauerstoffreduktion statt. Hieraus resultiert ein Absenken der Zellspannung um ca. 1 V. Unter den gegebenen Elektrolysebedingungen (80 °C) weist Sauerstoff jedoch nur eine geringe Löslichkeit im Elektrolyten (30 – 32 wt.% NaOH) auf. Daher ist es notwendig, die SVK als Gasdiffusionselektrode (GDE) auszulegen. Hierbei handelt es sich um poröse silberbasierte Elektroden, die durch den Einsatz von Polytetrafluorethylen (PTFE) hydrophobe Bereiche aufweisen. Im Betrieb dringt der flüssige Elektrolyt in die Porenstruktur ein, das PTFE verhindert aber ein komplettes Fluten der Elektrode. Über ein Gaskompartiment wird der Sauerstoff zugeführt und gelangt ebenfalls in das Innere der Elektrode. Es bildet sich eine Dreiphasengrenzfläche, bestehend aus flüssigem Elektrolyt, Gas und katalytisch aktivem Feststoff aus, an dem die elektrochemische Reaktion stattfinden kann. Obwohl die Technologie bereits erfolgreich industriell eingesetzt wird, sind viele Prozesse, insbesondere die Elektrolytverteilung, im Inneren der GDE weiterhin unbekannt.

Um weitere Einblicke in die Funktionsweise der GDE zu erhalten wurde bei der Herstellung der Anteil des PTFEs systematisch variiert und er Einfluss sowohl auf das Porensystem als auch auf die elektrochemische Leistung mittels Halbzellversuchen untersucht. Hierdurch lassen sich Hypothesen über den Grad der Flutung und damit auf die Elektrolytverteilung aufstellen. Dies liefert jedoch auch nur einen begrenzten Einblick. Über Radiographien mittels Synchrotronstrahlung während des Betriebs in der Halbzelle konnte erstmals eine Operando Elektrolytverteilung für silberbasierte GDE ermittelt werden. Durch die Implementierung dieser in ein verbessertes sogenanntes thin-film flooded agglomerate (TFFA) Modell konnten die Prozesse innerhalb der Elektrode analysiert werden. Dies gewährt detaillierte Einblicke in die Prozesse, z.B. wurde ein komplexer Wasserkreislauf identifiziert, durch den Hydroxidionen aus der GDE entfernt werden, womit die Leistung der Elektrode verbessert wird. Die Eindringtiefe des Elektrolyten ist allerdings bisher nicht durch Operando Experimente zu ermitteln. Dafür wurden in weiterführenden Experimenten gradierte Elektroden hergestellt. Durch die Verwendung von inaktiven Diffusionsschichten konnten die reaktiven Zonen der GDE eingegrenzt und so die effektive Eindringtiefe des Elektrolyten ermittelt werden. Hierdurch konnte aufgezeigt werden, dass sehr viel größere Regionen der Elektrode aktiv sein müssen, als bisher angenommen wurde.

Acknowledgements

I would like to express my sincere gratitude to all the people who supported me in writing this thesis.

First of all, I would like to thank my supervisor Prof. Dr.-Ing. Thomas Turek, who gave me the opportunity to work on such an exciting project. Not only the possibility to travel and to explore many different aspects of the topic, but also the countless discussions and suggestions were the reason to bring me scientifically, but also personally steadily one step further.

My thanks also go to Prof. Dr.-Ing. Ulrich Kunz for reviewing this thesis, but also to his almost immeasurable knowledge, through which he was always able to assist with any experimental questions.

I would like to thank Prof. Dr.-Ing. Ulrike Krewer from Karlsruhe Institute of Technology for taking over the review of my thesis. Also the many small conversations and discussions during project meetings and conferences were always an enrichment.

I thank Petra Ritter for her continuous support in all administrative matters.

A special thanks goes to Barbara Ellendorff. Without your support, this work would probably not have come to a successful conclusion.

This applies equally to the rest of the DFG research unit. The numerous meetings and discussions all contributed to the success of this thesis. Due to the relaxed atmosphere, they almost never felt like work. I would especially like to mention Max Röhe and our fruitful conversations about modeling. Marcus Gebhard for the exchanges about cell development and experimental design, and Melanie Paulisch who always provided us with exciting insights.

I would also like to thank my colleagues at the Institute. Mauritio Müller for the assistance with problems in the lab, Thorben Muddemann for the many discussions across the screen, and not to forget Steffen Flaischlen, who is always up for a joke and keeps the good mood at the institute going.

I would also like to thank my student Christian for his support during the experiments.

A special thanks goes to my dissertation support group, which always helped to get one step further. A big thanks goes to Laurens Reining for starting it and Katharina Schafner for accommodating me. You guys showed me the way, especially in the early stages. And my gratitude also goes to Jan Martin and Eva Prumbohm, who always provided constructive input later on.

I would also like to thank my family. Without your support all this would not have been possible in the first place.

And most importantly, of course, to my girlfriend Isabelle Kroner, who has always lovingly supported, challenged, encouraged and, above all, endured me. I certainly couldn't have done it without you.

Content

1 Introduction

Modern energy politics is driven by the energy turnaround reducing the amount of fossil fuels. Not only the expansion of renewable energies, but also the general reduction in energy consumption is crucial to achieve the Paris Agreement goals [1,2]. For the chemical industry, chlorine is one of the most important educts for over 60 % of all chemical products [3]. Most of the chlorine is produced in the energy intensive chlor-alkali electrolysis. In 2017 approx. 89 million metric tons of chlorine gas were produced with an energy demand of 2.5 – 3.5 MWh per ton of chlorine [4]. In total, the electrical energy demand of the chlorine production is equal to approx. 3 % of the world wide industrial electrical energy consumption [5]. In the state-of-the-art membrane process chlorine gas is produced at the anode, while hydrogen evolves at the cathode. At the first thought, this seems promising, since three valuable products, chlorine, sodium hydroxide and hydrogen, are produced in a single process. However, in reality 10 – 15 % of the evolved hydrogen is not used at all, which corresponds to the production value of a 200 MW water electrolyzer [6]. Thermal utilization or conversion back into electricity by fuel cells is not economical [7]. By the introduction of the oxygen depolarized cathode (ODC) to the system, electrical energy savings of up to 25 % were achieved [8]. Instead of evolving hydrogen, oxygen is reduced at the cathode. The byproduct, sodium hydroxide, is not affected by the usage of the new technology.

Figure 1.1 shows the comparison of the membrane (left) and ODC (right) processes schematically. In both cases the anolyte consists of saturated brine (NaCl solution) and chlorine evolves at the anode according to eq. 1.1. The positively charged sodium ions and water are able to pass the ion-selective membrane into the catholtye chamber. At the cathode the water is reduced to hydrogen and hydroxide ions according to eq. 1.2. As the negatively charged ions cannot pass the membrane, the product streams (chlorine, hydrogen and NaOH solution) leave the electrolyzer with high purity. Taking the standard potentials into account, at least a cell voltage of 2.19 V is required to run the electrolysis. Even after multiple iterations of improvements and minimizing all cell resistances, e.g. by zero-gap design [8], industrial cells still operate between 3 V [9] and 3.3 V [10] at 6 kA m^{-2}. Further significant reductions are only possible by changing the reaction at the cathode. By employing the ODC in the process, the evolution of hydrogen is suppressed and instead oxygen is reduced according to eq. 1.3. The amount of the produced chlorine and NaOH solution is not affected, but the required voltage decreases significantly. In theory, the cell voltage is reduced to a value of 0.96 V. Due to the low solubility of oxygen at the process conditions (80 °C, 30 – 32 wt.% NaOH) [11], the ODC needs to be designed as a gas-diffusion electrode (GDE). Within the half-cell the GDE separates an electrolyte and gas compartment as shown in Figure 1.1 on the right-hand side. In the electrolyte compartment the NaOH solution accumulates, as hydroxide ions are formed inside the ODC and sodium ions pass the membrane from the anode chamber. Oxygen is supplied in an overstoichiometric ratio reacting inside the ODC. The unreacted oxygen leaves the cell together with evaporated water from electrolyte chamber. Including all voltage loses, an industrial electrolyzer operates at 2.2 V and 4 kA m^{-2} [4].

The reactions in the process are as follows with the standard potentials (E^0) vs. the standard hydrogen electrode (SHE) according to Bratsch [12].

anode

$$2\,\mathrm{Cl^-} \rightarrow \mathrm{Cl_2} + 2\,\mathrm{e^-} \qquad E^0 = 1.36\text{ V vs. SHE} \tag{1.1}$$

cathode

membrane $$2\,\mathrm{H_2O} + 2\,\mathrm{e^-} \rightarrow \mathrm{H_2} + 2\,\mathrm{OH^-} \qquad E^0 = -0.83\text{ V vs. SHE} \tag{1.2}$$

ODC $$2\,\mathrm{H_2O} + \mathrm{O_2} + 4\,\mathrm{e^-} \rightarrow 4\,\mathrm{OH^-} \qquad E^0 = 0.40\text{ V vs. SHE} \tag{1.3}$$

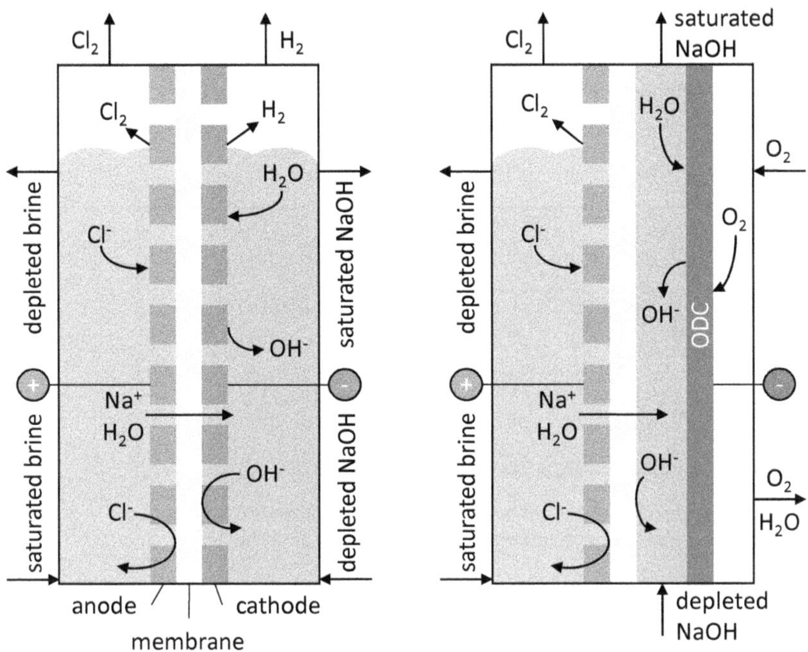

Figure 1.1: Comparison of the state-of-the-art membrane (left) and ODC (right) chlor-alkali electrolysis process. Adapted from [13].

The ODC is a complex structure with extremely high requirements, such as chemical and mechanical stability during operation, high surface area and activity of the electrocatalyst, and easy access of liquid and gaseous species without breakthrough of gas or flooding by the liquid electrolyte [14]. For the oxygen reduction reaction (ORR) silver serves as the electrocatalyst, as the activity is comparable to platinum at the given process condition [15]. The catalyst is mixed with a hydrophobic binder (most commonly polytetrafluoroethylene (PTFE)), which not only increases the mechanical stability, but also provides enough hydrophobicity to prevent the electrode from complete flooding. However, the electrolyte is still able to intrude the GDE at the more hydrophilic parts. A complex electrolyte distribution inside the GDE is created, forming the three-phase boundary and thus the available surface area for the electrochemical reaction. In Figure 1.2 a schematic overview of a single pore inside the GDE is given. Inside the pore the oxygen and the liquid electrolyte form a contact line, resulting in a meniscus of the electrolyte on the wall. In this region the electrolyte film is thin enough for oxygen to dissolve, diffuse to the catalyst and react together with the water to hydroxide ions. This limits the electrochemical active area indicated by the gray shaded area.

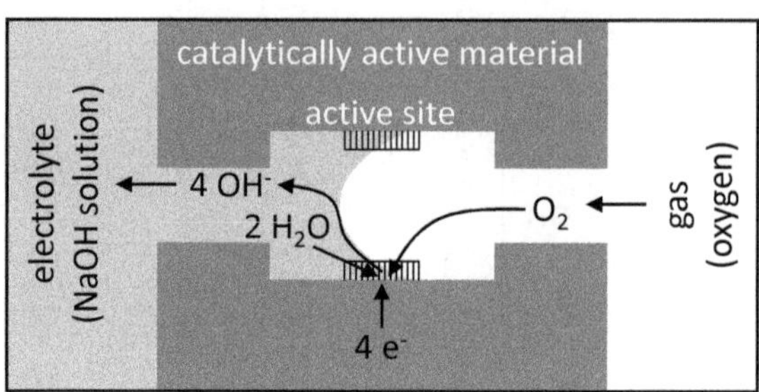

Figure 1.2: Model illustration of the three-phase boundary with transport and reaction phenomena in a single pore inside the ODC. Adapted from [13].

However, this is only a simple model conception. In reality, the structures inside the GDE a far more complex. Paulisch et al. [16] studied the pore system utilizing X-ray and synchrotron radiography, synchrotron tomography as well as focused ion beam milling and scanning electron microscopy (FIB/SEM) tomography. These techniques revealed an inhomogeneous system consisting of bigger pores connected to a complex homogeneous pore network microstructure. Examples of the FIB/SEM images are also shown in chapter 3. Neumann et al. [17] presented a pluri-Gaussian model describing the silver framework and the PTFE deposited in the pore space based on a FIB/SEM tomography. Such stochastic models allow the creation of realistic geometries in large scales, e.g. for further modeling. However, looking at Figure 1.2, it is clear, that not only the morphology of the pore system is crucial, but also the accessibility of both, liquid electrolyte and gaseous species. Therefore Kunz et al. simulated the imbibition of the electrolyte on the pore scale in 2D [18] and 3D [19]. However, their simulations require great computational resources and are only carried out on smaller scales. Coupling their simulations with electrochemical reactions would be even more challenging.

In general, for the modelling the scope of application is of fundamental importance. In their review, Kubannek et al. [20] summarized studies on modeling the ORR in GDE through all scales. For example, Wang [21] is able to describe the morphology changes of an ODC in lithium-air batteries during discharge with a 0D model. Röhe et al. [22,23] developed a 1D model of the ODC in the chlor-alkali electrolysis focusing on the dynamic processes, while Sijabat et al. [24] described the ion transport through the membrane with a 1D model. Vasile et al. [25] focused in the distribution of oxygen near the ODC surface in proton-exchange membrane fuel cells and utilized a 3D model for that. The possibilities are endless, but a tradeoff must always be made between available computational resources, level of detail and the numbers of parameters to be determined.

This thesis focuses only on the processes inside the GDE during ORR in highly alkaline electrolyte. Already in 1965 Austin et al. [26] developed a mathematical description of electrochemical reactions in GDE based on a model system as illustrated in Figure 1.2. The gas dissolves in the electrolyte at the meniscus, diffuses to the electrocatalyst and reacts on the active side. However, only very small three-phase interfaces, and thus small limiting current densities, are achieved in this way. In order to extend the available reaction surface, e.g. Srinivasan et al. [27], Iczkowski [28] and Will [29,30] enlarged the meniscus changing it into a thin-film laying on the whole pore surface. In contrast to that, e.g. Burshtein et al. [31] and Giner et al. [32] divided the electrode in hydrophilic and hydrophobic areas. Parts of the electrode are flooded, forming the so-called flooded agglomerates consisting of electrocatalyst and electrolyte, while the other parts of the electrode transport the gaseous species to the reaction sides. In 1975 Cutlip [33] combined both model approaches developing the so-called thin-film flooded agglomerate (TFFA) model. The electrode consist of hydrophobic gas channels and flooded agglomerates covered with a thin-film of electrolyte. The gaseous species diffuses through the gas channels, dissolves at the thin-film interface and diffuses to the flooded agglomerate, where the reaction takes place. Wang and Koda [34,35] adapted the model to the ORR for the chlor-alkali electrolysis. The electrolyte intrude the GDE in a finger shaped way, forming the flooded agglomerates. In between these electrolyte fingers the gaseous oxygen is able to diffuse to the reaction side. Later, Pinnow et al. [36] extended the model mainly by considering binary Maxwell-Stefan diffusion in the gas and electrolyte phase and validated it with half-cell polarization curves. However, recent studies by Botz et al. [37] showed, that the electrolyte transport in the GDE was underestimated. Further, the electrolyte distribution was unknown, due to the lack of operando experiments at this time. Thus, the geometric parameter were adjusted to match the half-cell experiments. In this thesis, the model of Pinnow et al. [36] is extended progressively to eliminate the unknown parameters. In the end, the improved model describes the processes inside the GDE even better and will give further insights.

The present thesis is structured into published manuscripts, which each contribute to the overall topic as follows:

- **Chapter 3** describes the influence of the PTFE content on the GDE. The techniques to analyze the pore system and the electrochemical performance via half-cell experiments are introduced. A commercially available half-cell and an in-house test set-up, designed for stationary conditions at the GDE, are utilized. The latter allows the application of pressure gradients between electrolyte and gas compartment. The experiments show that the electrolyte distribution is of fundamental importance, but cannot give any information of the shape and intrusion depth of the electrolyte.
- **Chapter 4** describes the development of the improved TFFA-model. The electrolyte distribution is determined independently from half-cell polarization curves during synchrotron operando experiments. Further, the electrolyte transport is considered adequately and all relevant physical parameters are integrated as correlation equations in dependence of the actual electrolyte concentration inside the GDE. As a result, the polarization curves can be predicted over a wide range of oxygen and electrolyte concentrations, while reducing the amount of unknown fitting parameter.
- **Chapter 5** addresses the question of whether the intrusion depth of the electrolyte can be determined. Graded electrodes, consisting of nickel diffusion layers and silver reaction layers, were manufactured according to the assumed subdivision of the TFFA-model. In this way, the effective

intrusion depth of the electrolyte is determined experimentally and compared with the simulation results. This gives further insights on the actual electrolyte distribution.

2 Overview of the publications

The following publications were submitted and published in "peer-reviewed" scientific journals and are an integral part of this doctoral thesis.

1. D. Franzen, B. Ellendorff, M. C. Paulisch, A. Hilger, M. Osenberg, I. Manke, T. Turek; Influence of binder content in silver-based gas diffusion electrodes on pore system and electrochemical performance, J. Appl. Electrochem., 49, 705–713 (2019).

   The actual impact factor of the Journal of Applied Electrochemistry is 2.398 (2019). The contribution of the first author was to design and conduct the electrochemical experiments and to prepare the manuscript. B. Ellendorff manufactured and physically characterized the Electrodes. M. C. Paulisch, A. Hilger and M. Osenberg executed FIB/SEM-tomography and reconstruction supervised by I. Manke. T. Turek advised in the experiment conception and contributed to writing the manuscript. This publication is reprinted in chapter 3 and describes the influence of the binder content on the pore system and the electrochemical performance. The utilization of the pore system and thus possible electrolyte distributions are discussed. Additionally, the experimental setup as well as physical characterization techniques are described in detail.

2. D. Franzen, M. C. Paulisch, B. Ellendorff, I. Manke, T. Turek; Spatially resolved model of oxygen reduction reaction in silver-based porous gas-diffusion electrodes based on operando measurements, Electrochim. Acta, 375 (2021) 137976.

   The actual impact factor of the Journal Electrochimica Acta is 6.216 (2020). The contribution of the first author was to design and conduct the electrochemical experiments, develop the mathematical model and to prepare the manuscript. M. C. Paulisch conducted and evaluated the operando experiments as well as supported to prepare parts of the manuscript concerning the operando experiments supervised by I. Manke. B. Ellendorff manufactured the Electrodes. T. Turek advised in the experiment and model conception and contributed to writing the manuscript. This publication is reprinted in chapter 4 and describes the development of the improved TFFA-model. The shape of the electrolyte distribution is implemented on the basis of the first available operando synchrotron experiments. Further, the electrolyte properties are spatially resolved, allowing a description of all processes for different NaOH and oxygen concentrations.

3. D. Franzen, C. Krause, T. Turek; Experimental and model-based analysis of electrolyte intrusion depth in silver-based gas diffusion electrodes, ChemElectroChem, 8, 2186–2192 (2021).

   The actual impact factor of the Journal ChemElectroChem is 4.154 (2019). The contribution of the first author was to design and evaluate the experiments, as well as to adapt the mathematical model. C. Krause prepared the electrodes and conducted the experiments. T. Turek advised in the experiment and model conception, and contributed to writing the manuscript. This publication is reprinted in chapter 5 and describes the influence of the electrolyte intrusion depth on the electrochemical performance. The reaction zone is precisely defined by specially designed graded electrodes. The geometrical parameters of the graded GDE can be transferred to the previously provided TFFA-model, resulting in a model-supported analysis of the effective penetration depth of the electrolyte.

3 Influence of binder content in silver-based gas diffusion electrodes on pore system and electrochemical performance

Reproduced and adapted under the creative common (CC BY 4.0) license.

D. Franzen, B. Ellendorff, M. C. Paulisch, A. Hilger, M. Osenberg, I. Manke, T. Turek
J. Appl. Electrochem., 49, 705–713 (2019)
https://doi.org/10.1007/s10800-019-01311-4

Graphical abstract

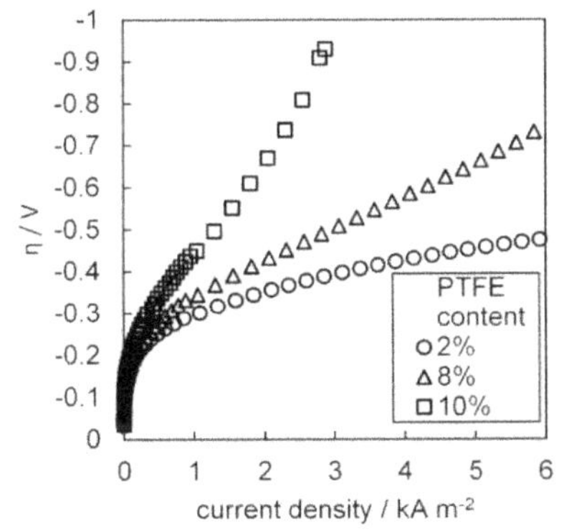

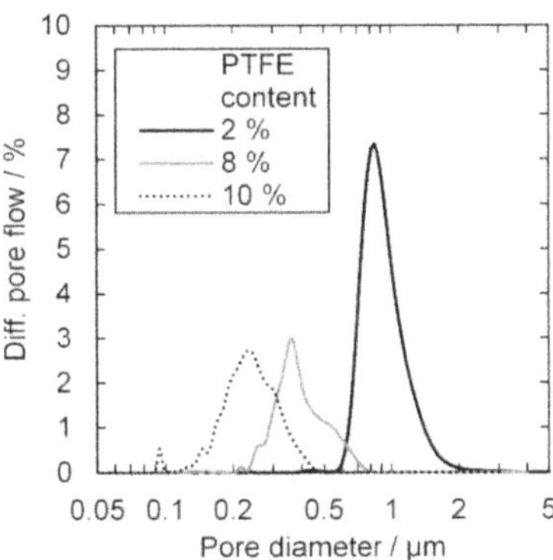

Abstract

The influence of the polytetrafluoroethylene (PTFE) content in silver-based gas diffusion electrodes on the resulting physical properties and the electrochemical performance during oxygen reduction in concentrated sodium hydroxide electrolyte were investigated through half-cell measurements. A systematic variation of the pore system was achieved by application of different silver/PTFE ratios during the production of the gas diffusion electrodes (GDE). In all electrodes, a silver skeleton structure with relatively constant properties was formed while the PTFE fills up part of the open pore space. The resulting structures were characterized with a variety of methods for the physical properties supported by focused ion beam milling and scanning electron microscope (FIB/SEM) tomography. It could be shown that variations in the obtained pore system strongly influence the electrochemical performance of the electrodes. Determination of the Tafel slopes revealed that this is not due to changes in the electrocatalytic activity but rather caused by variations in the electrolyte uptake. While too small amounts of PTFE (1 wt%) lead to decreased performance through electrolyte flooding, higher PTFE contents above about 5 wt% also deteriorate the electrode performance because the extent of the three-phase boundary diminishes. The decisive role of the electrolyte intrusion was confirmed by measurements at higher electrolyte pressure. While the best electrochemical performance was achieved with an electrode containing 98 wt % silver, a slightly higher PTFE content is advisable to prevent breakthrough of the electrolyte.

Keywords: oxygen reduction reaction, gas diffusion electrode, chlor-alkali electrolysis, oxygen depolarized cathode, silver

3.1 Introduction

The global chlorine production is mainly based on the chlor-alkali electrolysis, which is one of the most energy intensive processes in the chemical industry [1]. During the classical variant of the electrolysis process, chlorine gas is formed at the anode, while hydroxide ions are produced together with hydrogen gas at the cathode. A significant reduction of the electrical energy demand of this process was obtained by introducing a gas diffusion electrode (GDE) to the process resulting in oxygen reduction at the cathode rather than hydrogen evolution. The achievable cell voltage decreases of approximately one volt lead to electrical energy savings of up to 30 % of this so-called oxygen depolarized cathode (ODC) technology [2]. The development of processes and suitable catalyst materials for the required GDE have already been discussed in previous work [3]. Despite intensive research over several decades, the stability of carbon-based materials has proven to be insufficient under the harsh process conditions during technical electrolysis (80-90 °C, 30-32 wt% NaOH). On the other hand, silver has a similarly good activity for the oxygen reduction reaction (ORR) under these conditions [4]. Hence, commercial GDE for ODC electrolysis are carbon- and platinum-free and based on silver as electrocatalyst.

In previous work, a variation of silver raw materials and compositions of silver-based GDE lead to optimized electrodes regarding energy consumption and long-term stability [5]. However, proper structure-property relationships explaining why certain electrodes perform better than others are still lacking. Furthermore, models describing the processes within the GDE [6] and ODC chlor-alkali electrolysis cells [7] were developed. These simulations revealed that especially at high current densities, an insufficient supply of oxygen at the surface of the electrocatalyst occurs. That would mean that mass transport rather than the electrochemical activity is the limiting factor in these electrodes. This assumption is supported by findings obtained with electrodes containing very low amounts of the silver catalyst in dendritic form. These materials also revealed excellent performance, however at the expense of lower long-term stability [8]. It can be expected that a proper characterization of the pore system and the processes in the pores offer further major potential for improvement of silver-based GDE.

An efficient tool in the visualization of internal pore structures is the focused ion beam (FIB) tomography combined with a scanning electron microscope (SEM) [9]. On the basis of this image data, reconstructions of the pore system as 3D models coupled with numerical simulations of the transport processes are possible [10]. Disadvantageous is, the provided image sections are rather small and offer only a limited overview. For the GDE examined in this work a stochastic model based on FIB tomography was developed that represents the structure on a larger scale [11]. The structure serves as a basis for further numerical simulations considering e.g. transport phenomena inside the electrode. However, such large-scale simulations still pose challenges to modern computing systems.

Therefore, in this contribution the physical and electrochemical properties of silver-based GDE are analyzed using established characterization methods

[12] supported by FIB tomography and the generated 3D reconstructions [13]. For a series of electrodes with different composition, the resulting electrochemical performance is determined via half-cell measurements.

3.2 Experimental

3.2.1 Electrode preparation

For each electrode a suspension containing 30 g silver particles (SF9ED, Ferro GmbH) and a PTFE dispersion (TF 5060GZ, 3M™ Dyneon™) in the desired ratio was prepared. 50 g of a methyl cellulose solution containing 1 wt% hydroxyethyl methyl cellulose (WALOCEL™ MKX 70000 PP 01) and demineralized water was added as pore building agent and thickener. Depending on the GDE composition, a further amount of demineralized water was required to maintain a proper viscosity. The suspension was applied on a nickel mesh as conductive supporting material (106 µm x 118 µm mesh size, 63 µm thickness, Haver & Boecker OHG) using a spraying piston (Evolution, 0.6 mm pin hole, Harder & Steenbeck). For each electrode 80 coatings were applied on a heating table allowing simultaneous drying leading to the formation of a homogeneous surface [14]. In the subsequent processing steps the electrode was hot pressed (LaboPress P200S, Vogt, 15 MPa, 130 °C, 5 minutes) and heat-treated in an air oven (330 °C, 15 minutes) to burn out the methylcellulose and to improve the mechanical stability through PTFE sintering. Afterwards, the electrode thickness and catalyst load were determined. The mean thickness was obtained from measurements at six points with a thickness dial gauge (FD 50, Käfer GmbH), while the catalyst load was determined by weighing. For industrial applications silver contents of 92 to 98 wt% facing the electrolyte side and 95 to 99.9 wt% facing the gas side respectively with catalyst loadings of 140 to 250 mg cm^{-2} are desired [15]. To investigate a uniform porous layer, only single-layer GDE with silver contents of 90 to 99 wt% were prepared. A list of the properties for all electrodes produced is provided in Table 3.1. In the following the electrodes will be referred to by values for the silver content rounded to full weight percentage points.

Table 3.1: Overview about electrode properties.

Silver content	Silver load	PTFE load	thickness
wt%	mg cm^{-2}	mg cm^{-2}	µm
99.0	155.7	1.7	356
97.9	150.5	3.2	335
97.0	131.4	4.1	297
96.0	129.1	5.4	293
94.9	125.6	6.8	292
94.0	131.6	8.4	302
93.0	139.9	10.5	321
91.8	138.3	12.3	319
90.8	158.9	16.1	375
89.9	136.9	15.4	330

3.2.2 Physical characterization

The true density of the produced electrodes was obtained with a helium pycnometer (Pycnomatic ATC, Quantachrome), which measures the volume of the displaced helium. As the helium penetrates all open pores the true density is independent of the porosity. On the other hand, the expected density was calculated based on the electrode weight and the true densities of silver (10.49 g cm^{-3}), PTFE (2.2 g cm^{-3}) and nickel (8.91 g cm^{-3}), and should also be independent of the porosity. Flow-through pores and bubble point pressure were determined using capillary flow porometry (Porometer 3G, Quantachrome). A wetting fluid (Porofil, Quantachrome) was applied to the probe which was adjusted in the device. Afterwards, the fluid was driven out of the pores with a pressure gradient. The resulting nitrogen flow was detected on top of the probe. The bubble point pressure was determined at a nitrogen flow rate of 0.1 L min^{-1}. The flow-through pore distribution was finally determined by comparing the nitrogen flow through the wet and dry probe. Specific surface areas of the electrodes were determined using the BET method (3 Flex, Micromeritics Instrument Corp.) with krypton as a sample gas. This minimized the device error in view of the small overall inner surface area. Mercury porosimetry (Pascal 140+440, Thermo Fisher Scientific) measurements were tested as an additional method, but due to amalgam formation, reproducible results could not be obtained. Visualization of the microstructure inside the electrodes was investigated by means of FIB milling combined with SEM. Segments with a size of 2 mm x 4 mm were extracted mechanically from the center of the GDE and fixed on SEM sample holders with carbon pads. For ion cutting a Zeiss Crossbeam 340 Gallium-FIB/SEM device was used. The coarse cut was done using an acceleration voltage of 30 keV and an ion current of 50 nA. For the final polishing setup 30 keV and 700 pA were employed. The silver fraction in the GDE microstructure was calculated with the imaging software Fiji [16].

3.2.3 Electrochemical characterization

Electrochemical properties were determined in half-cell measurements (FlexCell HZ-PP01, Gaskatel GmbH) at 80 °C using 30 wt% NaOH as electrolyte prepared from caustic flakes (≥99 wt%, Carl Roth) and demineralized water. All reported current densities are related to the geometric cell area of 3.14 cm^2. Experiments were performed with a Zennium Pro Potentiostat (Zahner GmbH). The electrodes were characterized following the same routine, including a start-up procedure and cell resistance determination with pseudo-galvanostatic impedance measurements. Finally, an iR compensated linear sweep voltammetry (LSV) measurement starting at open cell potential (OCP) to 200 mV vs. reversible hydrogen electrode (RHE) with a scan rate of 0.5 mV s^{-1} was performed.

The half-cell contained a gas and an electrolyte chamber separated by the GDE. Measurements were performed using a three-electrode configuration with the GDE as working electrode. As the counter electrode a platinum wire was placed inside the electrolyte chamber with a volume of approximately 30 mL. The potentials were measured in front of the working electrode using a Luggin capillary with a RHE (Hydroflex, Gaskatel GmbH). On the gas side pure oxygen was supplied with a flow rate of 50 mL_N min^{-1} and a small backpressure using a 1 mm water column.

Additional experiments applying an increased electrolyte pressure were performed in a special plant with an electrolyte cycle and separate pressure control for the electrolyte and gas supply. The half-cell design (Figure 3.1) is similar to the FlexCell, providing the same electrode configuration. Pressure and temperature on gas and electrolyte side were monitored with sensors. During these experiments gas and electrolyte pressure were set to 1000 mbar first. Subsequently, the electrolyte pressure was increased stepwise. All other process parameters were kept at the same conditions as during the experiments in the FlexCell.

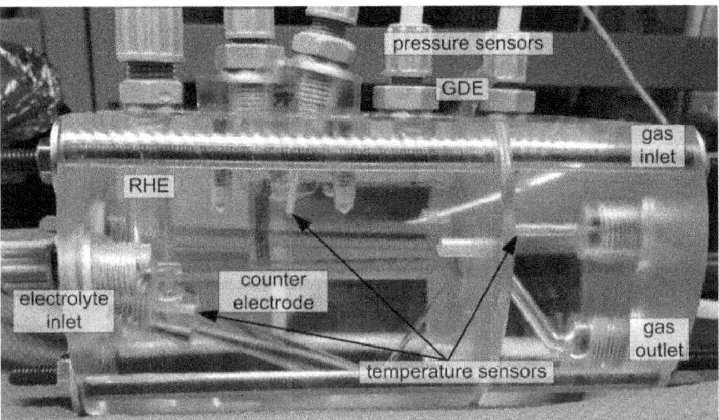

Figure 3.1: Half-cell for the measurements at elevated electrolyte pressure, electrolyte outlet is located behind the counter electrode.

To determine proper equilibrium potentials of the oxygen reduction reaction (ORR) vs. RHE, standard potentials for both reactions were calculated using the Nernst equation. The temperature dependence of the standard potentials was taken into account according to Bratsch [17], while the activity of water [18] and OH^- ions [19] were based on the work of Balej. Additionally partial pressures of hydrogen, at the RHE, and oxygen, at the GDE, were corrected by the water vapor pressure above the NaOH solution at the given temperature [18]. Finally the solubility of oxygen was determined with the model of Tromans [20]. Overall the standard potential of the ORR vs. RHE could be determined as 1.13 V at the given process conditions.

Evaluation of the LSV measurements was done by determining Tafel slopes and exchange current densities from the polarization curve. As will be shown later, two linear regions on the logarithmic current density scale were observed. This behavior is typical for the ORR in alkaline electrolyte [21] and was observed by Pinnow et al. [6] during their measurements with Ag/PTFE electrodes. While the first Tafel slope was obtained for overpotentials between 80 mV and 120 mV, the second slope refers to overpotentials in the range between 200 mV and 250 mV. For all measurements, the range was defined individually to maximize the linear correlation factor.

3.3 Results and discussion

3.3.1 Influence of PTFE content on pore system

In Figure 3.2 the results of the porometry and bubble point measurements are shown for all prepared electrodes. It can be seen that the electrode with the highest silver content (99 wt%) has the largest median pore size of about 1 µm and a relatively sharp pore size distribution. Increasing the PTFE content decreases the pore size and broadens the pore size distribution. For the lowest applied silver content of 90 wt% the median pore size is decreased to approx. 0.2 µm. The right hand diagram in Figure 3.2 shows that rising PTFE content and thus smaller pores lead to an increase of the bubble point pressure. While the bubble point pressure of the electrode with 99 wt% silver is only 0.5 bar, this value rises strongly reaching 2.5 bar for 90 wt% silver. However, these measurements were performed with a wetting fluid at room temperature and bubble points could differ when using the electrolyte at process conditions. Furthermore electro wetting effects are not considered [22]. Nevertheless, the bubble point pressure is a good indicator for the wettability at process conditions. A low value will probably lead to a complete flooding of the electrode and thus higher transport resistance through the electrolyte. High values could correspond to severe blocking of the electrolyte paths inside providing only minimal electrolyte penetration. In both cases the accessible reactive surface of the GDE would be strongly reduced.

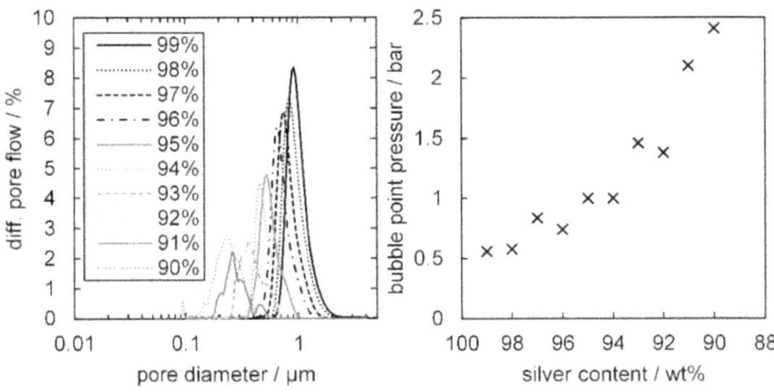

Figure 3.2: Results of capillary flow porometry and bubble point for electrodes with different silver content.

The FIB/SEM images (Figure 3.3) of the electrodes provide a look inside the electrodes, revealing details of the pore system. While the bright parts are silver, the darker shapes represent PTFE. The high contrast between these two phases allows for a clear differentiation. It can be seen that the electrode with 99 wt% silver has almost no PTFE inside the pores. With rising PTFE content of the electrode the silver surface becomes increasingly covered with PTFE. The investigated pores of the electrode with 91 wt% silver appear to be almost completely filled with the PTFE. Interestingly, the shape of the silver structure remains constant and is apparently not influenced by the amount of PTFE. Hence, the PTFE only fills up part of the free space inside the silver skeleton. According to the image analysis all electrodes consist of silver with approx. 60 % volume fraction, while 40 % are shared by open pore space and PTFE. However, it has to be kept in mind that the images show only a small area of the whole electrode. The results of the capillary flow porometry reveal that open pathways must exist even for electrodes with high amounts of PTFE. On the other hand, PTFE apparently blocks part of the open silver surface and the electrochemical activity of these electrodes is expected to drop significantly.

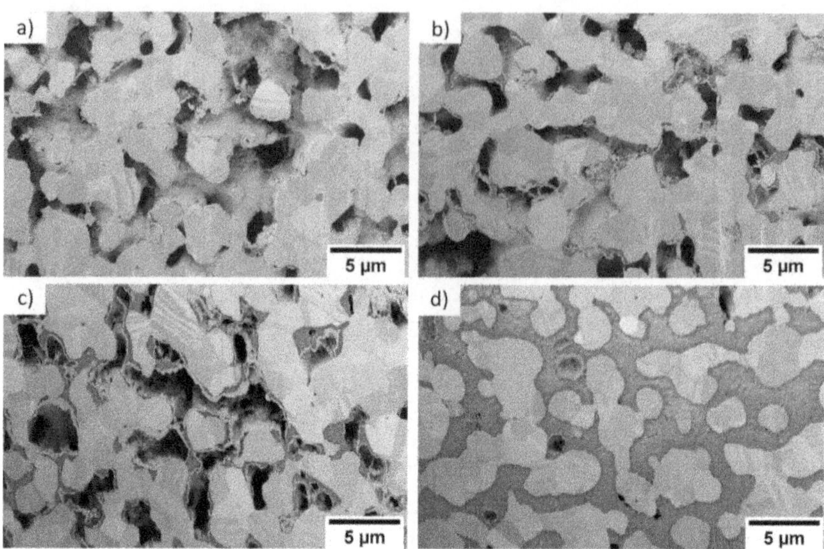

Figure 3.3: FIB/SEM images of four electrodes with 99 wt% (a), 98 wt% (b), 96 wt% (c) and 91 wt% (d) silver.

In Figure 3.4 true and expected electrode densities as well as the BET surface areas are shown. While the expected density, which is calculated based on the true material densities and the weight data, slightly decreases with rising PTFE content, a much stronger decline is observed for the true density determined by pycnometry. This result is an indicator for closed pores within the electrode which cannot be intruded by the helium during pycnometry. The total volume of the closed pores becomes larger with a decreasing amount of silver reaching approximately 20 % of the whole electrode volume at 90 wt% silver. As these closed pores should not take part in the electrochemical reaction, the silver catalyst utilization is expected to decrease with rising PTFE content.

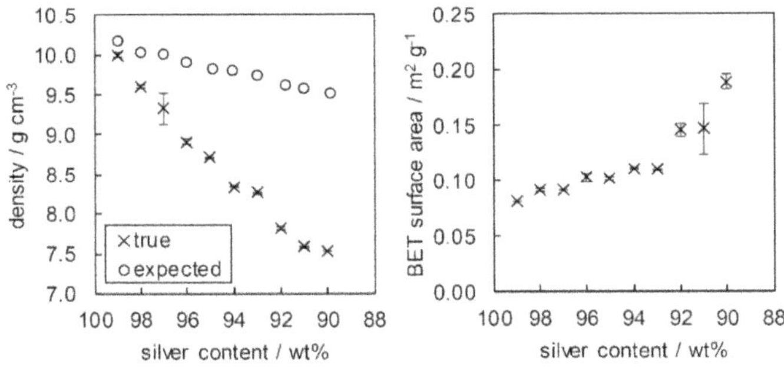

Figure 3.4: Densities and BET surface area as a function of silver content.

The right-hand diagram of Figure 3.4 shows that the PTFE content of the electrode does only have minor effects on the BET surface area which is in the range of 0.1 $m^2\ g^{-1}$. Only for low silver contents of 92 wt% and less the surface area slightly increases, however, at rising measuring errors. According to the manufacturer's specifications, the BET surface area of the pristine silver particles is between 0.7 and 1.2 $m^2\ g^{-1}$, while the surface area of the PTFE particles corresponds to approx. 13 $m^2\ g^{-1}$ derived from the given diameter of the particles. Obviously, the surface area of the starting materials decreases quite strongly during the fabrication process for which pore filling and sintering might be responsible.

3.3.2 Influence of PTFE content on ORR activity

The cell resistance for each electrode was determined using electrochemical impedance spectroscopy. The obtained values were in the range between 70 to 200 mOhm without any correlation to the silver content. It is therefore assumed that the conductivities of all electrodes are sufficient for a technical application. The iR-compensated polarization curves of the electrodes with different composition are shown in Figure 3.5 revealing a clear trend of the electrochemical performance in dependence of the silver content. While all electrodes exhibit nearly the same overpotential for low current densities, the differences are getting more and more pronounced for current densities above 1 kA m^{-2}. The electrodes with the smallest silver content of 90 wt%, 91 wt% and 92 wt% reach their limiting current density at 2 kA m^{-2}, 4 kA m^{-2} and 7 kA m^{-2}, respectively. On the other hand, the 99 wt% electrode does show transport limitations, too, but the effect is less significant. While the lower performance of the electrodes with high PTFE content might be explained by pore blocking and the resulting low filling degree with electrolyte, the high overpotentials for the 99 wt% electrode are most likely the result of a too low PTFE content which allows flooding with electrolyte. The lowest overpotentials are reached with a silver content of 98 wt%, which slightly differs from the optimum of 97 wt% found by Moussallem et al. [5]. Even for the highest applied current density of 8 kA m^{-2}, this electrode does not show any sign of transport limitations. The electrodes with silver contents between 97 wt% and 93 wt% do neither reach a limiting current density, but their overpotentials become slightly larger with decreasing silver content.

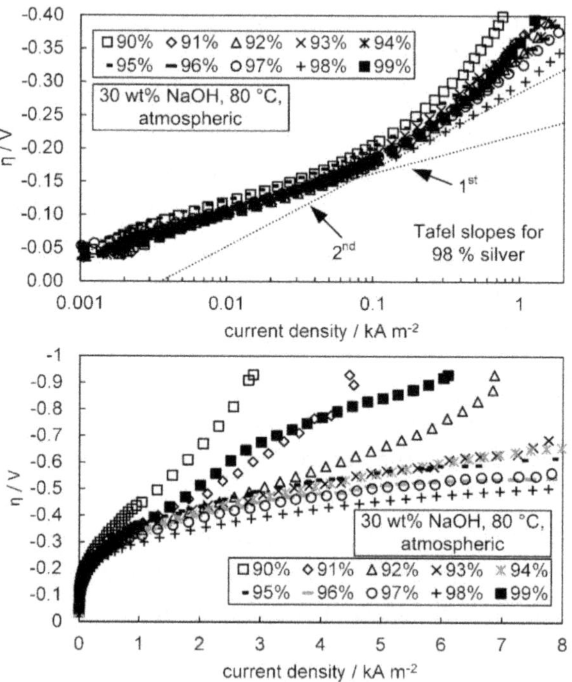

Figure 3.5: Overview of polarization curves for electrodes with different silver contents in logarithmic (top) and linear (bottom) representation.

The logarithmic diagram in Figure 3.5 reveals that the polarization curves can be divided into two regions with different Tafel slopes. The resulting values for the Tafel slopes are depicted in the left-hand diagram of Figure 3.6. It can be seen that the Tafel slopes for the 1st range are not dependent on the silver content and have a constant value of -60 mV dec^{-1}. In the 2nd range the slopes change as a function of the silver content in the electrode. The best electrode with 98 wt% silver shows a slope of -120 mV dec^{-1}, which slightly decreases for rising PTFE contents with a minimum of -140 mV dec^{-1} at a silver content of 94 wt%.

This observed doubling of the Tafel slope is typical for ORR kinetics in alkaline electrolyte and has been reported by other researchers, too. Sepa et al. observed Tafel slopes of -60 mV dec^{-1} and -120 mV dec^{-1} for the ORR in aqueous LiOH solution in a temperature range of 5 °C- 45 °C [21]. Similar values where observed for different systems using rotating disc electrodes (RDE) [23–25]. At conditions similar to those employed in this study Blizanac et al. observed slopes of approx. -70 mV dec^{-1} and -130 mV dec^{-1} (0.1 M KOH, 60 °C) on single crystal surfaces, which were found to be independent of the crystal structure [26]. Pinnow et al. observed values of -80 mV dec^{-1} and -200 mV dec^{-1} with silver/PTFE GDE [6]. In addition to these experimental findings, Shinagawa et al. showed that the two Tafel slopes can also be derived from the microkinetics of the ORR [27].

For the 98 % electrode Tafel slopes close to the true kinetics were determined and almost no influence of transport phenomena can be seen. For decreasing silver content, an increasingly larger transport effect is overlapping resulting in decreasing Tafel slopes. Additionally, the corresponding exchange current densities in both ranges are depicted in the right-hand diagram of Figure 3.6 showing only little variation for all tested electrodes. These results indicate that the electrocatalytic activity of the silver particles in the different electrodes is very similar and independent of the particular GDE composition. On the other hand, this means that the overall electrode performance is due to the extent of the three-phase boundary between gaseous oxygen, liquid electrolyte and solid catalyst, which is determined by the filling degree and the individual liquid pathways in the complex three-dimensional pore system. For the given electrolyte distribution, not only the transport of oxygen but also the local ion activity distribution are of decisive importance as recently shown by Botz et al. [28] through scanning electrochemical microscopy (SECM) measurements.

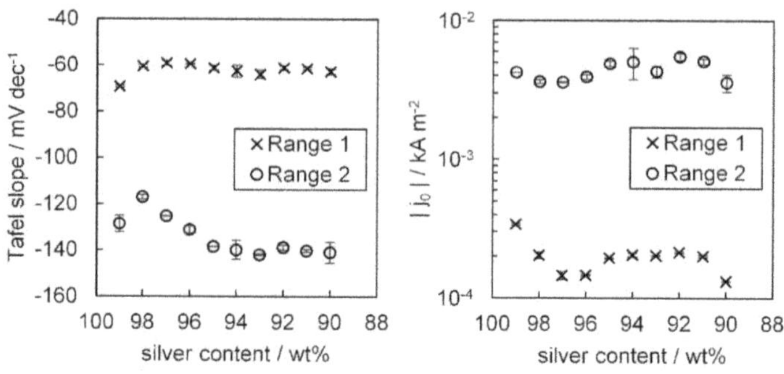

Figure 3.6: Tafel slopes (left) and exchange current densities (right) as a function of silver content.

3.3.3 Influence of electrolyte pressure on overpotentials during ORR

All experiments discussed so far were obtained at atmospheric pressure (approx. 950 mbar at Clausthal-Zellerfeld) with the Gaskatel half-cell. Selected electrodes with silver contents of 98 wt% and 94 wt% were additionally tested under various electrolyte pressure conditions in the specially designed half-cell. The results of these measurements are shown in Figure 3.7. It can be seen that for the best performing electrode with a silver content of 98 wt% an increased electrolyte pressure has a positive effect on the overpotentials for current densities below 3 kA m^{-2}. However, for a current density above approx. 5.5 kA m^{-2} the overpotential increases for higher electrolyte pressures. This result might be due to an increased filling level of the electrode pores with electrolyte resulting in deteriorated oxygen supply. This interpretation is supported by the fact that for the highest applied electrolyte pressure of 1200 mbar a larger amount of liquid passing from the electrolyte to the gas chamber was observed. For the 94 wt% electrode, a moderately increased electrolyte pressure of 1000 mbar has a clear positive effect on the electrochemical performance in the whole range of current densities. However, at even higher pressures a significant increase of the overvoltage at current densities above 5 kA m^{-2} becomes visible. At the highest pressure of 1200 mbar this effect becomes so strong that the electrode performance becomes worse than in the cell operated at atmospheric pressure. The increased pressure might lead to a deeper penetration of the electrolyte into the pore system and thus to a better performance. However, at too high pressures parts of the electrode are flooded and larger transport resistances in the electrolyte occur.

During technical application of the GDE in a chlor-alkali electrolysis cell, the electrode should be capable of withstanding a certain pressure difference between gas and electrolyte chamber without loss of performance. A stable electrolysis process is only guaranteed if the electrode is prevented from electrolyte breakthrough. Considering this, an electrode containing 97 wt% silver is still the best compromise between electrochemical performance and practical applicability [5].

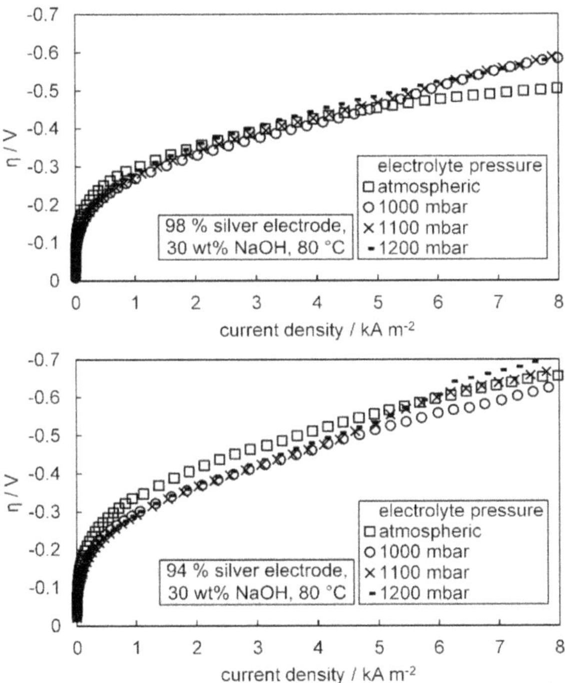

Figure 3.7: Polarization curves for two selected electrodes at different electrolyte pressures.

3.4 Conclusion

In the present contribution a systematic variation of the PTFE content in the production of silver-based GDE and a thorough physical and electrochemical characterization during oxygen reduction was carried out. Based on capillary flow porometry and FIB tomography it can be concluded that a silver skeleton is build up, while the PTFE is embedded into the free pore space and fills increasingly larger parts of the pore system. In addition, a rising fraction of completely inaccessible parts in the pore system is also formed at too high PTFE contents. Based on LSV measurements, a clear relationship between the formed pore system and the electrochemical performance was identified. On the one hand, all examined electrodes showed very similar electrocatalytic properties in the kinetic regime which can be described with two Tafel slopes showing the typical doubling characteristic for the ORR in alkaline media. On the other hand, the electrode performance at technically relevant higher current density is mainly determined by transport resistances. The best compromise between too strong filling with electrolyte at very low PTFE contents and too severe hydrophobic behavior at higher PTFE fractions offers an electrode containing 98 wt% silver. For this electrode, no signs of any transport limitation occur even at the highest applied current density of 10 kA m^{-2} under atmospheric electrolyte pressure. However, during application of an increased electrolyte pressure, signs of transport limitations at higher current densities emerged. Overall, the findings revealed that the electrolyte distribution inside the electrode is the key element for understanding the electrochemical performance. Therefore, in-situ measurements of a working silver-based GDE should be carried out revealing the electrolyte distribution, as it has been already done for carbon-based GDE used in fuel cells [29]. These measurements are presently carried out in our research groups and will be reported in a forthcoming paper.

3.5 Conflict of Interest

The authors declare that they have no conflict of interest.

3.6 Acknowledgments

Financial support for this study by Deutsche Forschungsgemeinschaft in the framework of the research unit "Multiscale analysis of complex three-phase systems: Oxygen reduction at gas-diffusion electrodes in aqueous electrolyte"(FOR 2397; research grants TU 89/13-1 and MA 5039/3-1). The authors also thank Holger Kropf for his support at the FIB instrument.

3.7 References

1. Kintrup J, Millaruelo M, Trieu V, Bulan A, Mojica ES (2017) Gas Diffusion Electrodes for Efficient Manufacturing of Chlorine and Other Chemicals. Electrochem Soc Interface 26(2): 73–76. doi: 10.1149/2.F07172if
2. Jörissen J, Turek T, Weber R (2011) Chlorherstellung mit Sauerstoffverzehrkathoden. Energieeinsparung bei der Elektrolyse. Chem unserer Zeit 45(3): 172–183. doi: 10.1002/ciuz.201100545
3. Moussallem I, Jörissen J, Kunz U, Pinnow S, Turek T (2008) Chlor-alkali electrolysis with oxygen depolarized cathodes: History, present status and future prospects. J Appl Electrochem 38(9): 1177–1194. doi: 10.1007/s10800-008-9556-9
4. Furuya N, Aikawa H (2000) Comparative study of oxygen cathodes loaded with Ag and Pt catalysts in chlor-alkali membrane cells. Electrochim Acta 45(25-26): 4251–4256. doi: 10.1016/S0013-4686(00)00557-0
5. Moussallem I, Pinnow S, Wagner N, Turek T (2012) Development of high-performance silver-based gas-diffusion electrodes for chlor-alkali electrolysis with oxygen depolarized cathodes. Chemical Engineering and Processing: Process Intensification 52: 125–131. doi: 10.1016/j.cep.2011.11.003
6. Pinnow S, Chavan N, Turek T (2011) Thin-film flooded agglomerate model for silver-based oxygen depolarized cathodes. J Appl Electrochem 41(9): 1053–1064. doi: 10.1007/s10800-011-0311-2
7. Chavan N, Pinnow S, Polcyn GD, Turek T (2015) Non-isothermal model for an industrial chlor-alkali cell with oxygen-depolarized cathode. J Appl Electrochem 45(8): 899–912. doi: 10.1007/s10800-015-0831-2
8. Frania P (2016) Herstellung, Analyse und Optimierung von Sauerstoffverzehrkathoden mit elektrochemisch abgeschiedenem Silberkatalysator zum Einsatz in der Chlor-Alkali-Elektrolyse. Dissertation, 1. Auflage. Technische Universität Dortmund; Verlag Dr. Hut
9. Zils S, Timpel M, Arlt T, Wolz A, Manke I, Roth C (2010) 3D Visualisation of PEMFC Electrode Structures Using FIB Nanotomography. Fuel Cells 10(6): 966–972. doi: 10.1002/fuce.201000133

10. Danner T, Eswara S, Schulz VP, Latz A (2016) Characterization of gas diffusion electrodes for metal-air batteries. J Power Sources 324: 646–656. doi: 10.1016/j.jpowsour.2016.05.108
11. Neumann M, Osenberg M, Hilger A, Franzen D, Turek T, Manke I, Schmidt V (2019) On a pluri-Gaussian model for three-phase microstructures, with applications to 3D image data of gas-diffusion electrodes. Comput Mater Sci 156: 325–331. doi: 10.1016/j.commatsci.2018.09.033
12. Arvay A, Yli-Rantala E, Liu C-H, Peng X-H, Koski P, Cindrella L, Kauranen P, Wilde PM, Kannan AM (2012) Characterization techniques for gas diffusion layers for proton exchange membrane fuel cells – A review. J Power Sources 213: 317–337. doi: 10.1016/j.jpowsour.2012.04.026
13. Salzer M, Spettl A, Stenzel O, Smått J-H, Lindén M, Manke I, Schmidt V (2012) A two-stage approach to the segmentation of FIB-SEM images of highly porous materials. Materials Characterization 69: 115–126. doi: 10.1016/j.matchar.2012.04.003
14. Moussallem I (2011) Development of Gas Diffusion Electrodes for a New Energy Saving Chlor-Alkali Electrolysis Process. Dissertation, Institute of Chemical Process Engineering, TU Clausthal
15. Turek T, Moussallem I, Bulan A et al. (2010) Oxygen-consuming electrode and method for its production(EP20110169579 20110610)
16. Schindelin J, Arganda-Carreras I, Frise E, Kaynig V, Longair M, Pietzsch T, Preibisch S, Rueden C, Saalfeld S, Schmid B, Tinevez J-Y, White DJ, Hartenstein V, Eliceiri K, Tomancak P, Cardona A (2012) Fiji: An open-source platform for biological-image analysis. Nat Methods 9: 676 EP -. doi: 10.1038/nmeth.2019
17. Bratsch SG (1989) Standard Electrode Potentials and Temperature Coefficients in Water at 298.15 K. J Phys Chem Ref Data 18(1): 1–21. doi: 10.1063/1.555839
18. Balej J (1985) Water vapour partial pressures and water activities in potassium and sodium hydroxide solutions over wide concentration and temperature ranges. Int J Hydrog Energy 10(4): 233–243. doi: 10.1016/0360-3199(85)90093-X
19. Balej J (1996) Activity Coefficients of Aqueous Solutions of NaOH and KOH in Wide Concentration and Temperature Ranges. Collect Czech Chem Commun 61(11): 1549–1562. doi: 10.1135/cccc19961549
20. Tromans D (2000) Modeling Oxygen Solubility in Water and Electrolyte Solutions. Ind Eng Chem Res 39(3): 805–812. doi: 10.1021/ie990577t

21. Sepa DB, Vojnovic MV, Vracar LM, Damjanovic A (1986) Apparent enthalpies of activation of electrodic oxygen reduction at platinum in different current density regions—II. Alkaline solution. Electrochim Acta 31(1): 97–101. doi: 10.1016/0013-4686(86)80068-8
22. Mugele F, Baret J-C (2005) Electrowetting: From basics to applications. J Phys Condens Matter 17(28): R705-R774. doi: 10.1088/0953-8984/17/28/R01
23. Blizanac BB, Ross PN, Markovic NM (2007) Oxygen electroreduction on Ag(111): The pH effect. Electrochim Acta 52(6): 2264–2271. doi: 10.1016/j.electacta.2006.06.047
24. Obradović MD, Grgur BN, Gojković SL, Vračar LM (2006) Enhancement of the electrochemical reduction of oxygen at platinum by nickel underpotential deposition. J Solid State Electrochem 11(1): 77–83. doi: 10.1007/s10008-005-0072-0
25. Chatenet M, Genies-Bultel L, Aurousseau M, Durand R, Andolfatto F (2002) Oxygen reduction on silver catalysts in solutions containing various concentrations of sodium hydroxide - comparison with platinum. J Appl Electrochem 32(10): 1131–1140. doi: 10.1023/A:1021231503922
26. Blizanac BB, Ross PN, Marković NM (2006) Oxygen reduction on silver low-index single-crystal surfaces in alkaline solution: Rotating ring disk(Ag(hkl)) studies. J Phys Chem B 110(10): 4735–4741. doi: 10.1021/jp056050d
27. Shinagawa T, Garcia-Esparza AT, Takanabe K (2015) Insight on Tafel slopes from a microkinetic analysis of aqueous electrocatalysis for energy conversion. Sci Rep 5: 13801. doi: 10.1038/srep13801
28. Botz A, Clausmeyer J, Öhl D, Tarnev T, Franzen D, Turek T, Schuhmann W (2018) Local Activities of Hydroxide and Water Determine the Operation of Silver-Based Oxygen Depolarized Cathodes. Angew Chem Int Ed 57(38): 12285–12289. doi: 10.1002/anie.201807798
29. Muirhead D, Banerjee R, George MG, Ge N, Shrestha P, Liu H, Lee J, Bazylak A (2018) Liquid water saturation and oxygen transport resistance in polymer electrolyte membrane fuel cell gas diffusion layers. Electrochim Acta 274: 250–265. doi: 10.1016/j.electacta.2018.04.050

4 Spatially resolved model of oxygen reduction reaction in silver-based porous gas-diffusion electrodes based on operando measurements

Reproduced and adapted under the creative common (CC BY 4.0) license.

D. Franzen, M. C. Paulisch, B. Ellendorff, I. Manke, T. Turek;
Electrochimica Acta, 375 (2021), 137976
https://doi.org/10.1016/j.electacta.2021.137976

Highlights

- Improved spatially resolved model for gas-diffusion electrodes with liquid electrolyte
- First operando electrolyte distribution measurements
- Description of electrode performance at different electrolyte and oxygen concentrations
- Model-based analysis of electrode processes reveals important role of water transport

Graphical abstract

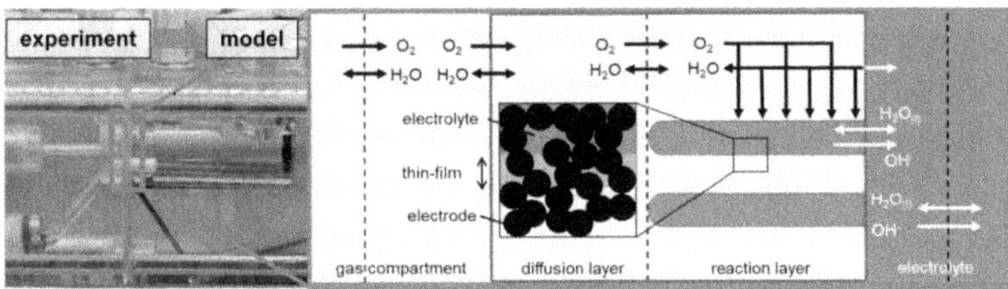

Abstract

An improved spatially resolved mathematical model for porous, silver-based gas-diffusion electrodes for the oxygen reduction reaction in strongly alkaline electrolyte is presented. For the first time, important parameters describing the distribution of the liquid electrolyte in the electrode were determined via independent operando synchrotron experiments. With the model, a reasonable description of the overvoltage as a function of the current density at 80 °C could be achieved in a range of electrolyte concentrations (27 – 36.6 wt.% NaOH) and oxygen contents (20 – 100 vol.%). Model-based analysis of the processes inside the electrode revealed that a complex water cycle evolves which supports the removal of hydroxide ions and thus boosts the electrode performance. The analysis also showed that additional operando measurement of the electrolyte distribution and investigations with electrodes having a defined thickness of the reaction layer will help to further improve the model.

Keywords: gas-diffusion electrode, silver, oxygen reduction reaction, modelling, operando measurements.

4.1 Introduction

Chlorine is one of the most important basic chemicals produced in large amounts by energy intensive electrolysis processes [1]. Through the introduction of the so-called oxygen depolarized cathode (ODC), which consumes oxygen instead of evolving hydrogen, electrical energy savings of up to 25 % were reached on an industrial scale [2]. Due to the low solubility of oxygen, the ODC is designed as a gas-diffusion electrode (GDE). Technically used GDEs contain silver as electrocatalyst, as under the harsh process conditions (30 – 32 wt.% NaOH, 80 °C) the oxygen reduction reaction (ORR) activity on silver is comparable to platinum [3]. Added polytetrafluoroethylene (PTFE) provides enough hydrophobicity to prevent a complete flooding of the GDE resulting in a high three-phase boundary between liquid electrolyte and gas inside the electrode. Although the ODC has been already introduced successfully in industrial applications, most of the details of the complex processes inside the electrode are still unknown. In previous studies, the interaction between inner geometry, composition and electrode performance was experimentally determined [4,5]. Moreover, it could be shown that modifications during the GDE manufacturing process improved the electrode performance [6]. In other approaches the required silver content was significantly reduced by selective deposition inside PTFE membranes according to a needle shaped electrolyte distribution [7]. However, this electrode type lacked long-term stability although the initial performance was comparable to industrial GDEs. All findings combined indicate that not only the electrocatalyst, but the pore system and the resulting electrolyte distribution inside the GDE are of crucial importance. Operando experiments could give insights, but the silver-based pore system is extremely challenging. New studies using micro-computed tomography (μCT) as radiation source have only recently become available [8,9]. However, due to the low resolution of the μCT, no exact predictions of the electrolyte distribution can be derived from these data. The modelling of the electrolyte distribution is currently only possible on very small scales with a high calculation effort. Kuntz et al. [10] predicted that it is necessary to apply high electrolyte overpressures to intrude the pore system, which is in contrast to the experimental studies where electrolyte even reaches the gas compartment under normal operation conditions [4,6,9,11].

On the basis of the work of Cutlip [12] and Wang and Koda [13,14], Pinnow et al. [15] developed a pseudo-2D so called thin-film flooded agglomerate (TFFA) model taking physical transport processes and electrochemical reaction into account. However, due to the lack of operando experiments at this time, the geometric parameters describing the electrolyte were adjusted to match the experimental results. The simulations predicted almost constant electrolyte concentrations inside the electrode and in the electrolyte bulk. In contrast, Botz et al. [16] showed a drastic increase of the local hydroxide ion concentrations in the electrolyte bulk near the GDE surface. Based on their findings, even higher concentrations inside the GDE must be expected. The changing local electrolyte concentration has also an influence on the oxygen solubility, which could not be captured with the model of Pinnow et al. either

[15]. An improved one-dimensional dynamic modelling approach by Röhe et al. [17,18] has already successfully demonstrated the importance of local electrolyte activities on the overall performance of the GDE. However, the spatial electrolyte distribution was not considered in this model. In another three-dimensional model approach by Vasile et al. [19] focusing on the oxygen transport through a gas-diffusion layer, only a simplified enhancement factor was used to describe the electrochemical reaction instead of an adequate description of the electrolyte distribution inside the catalytically active layer. Furthermore, there is no systematic study so far describing the influence of oxygen and electrolyte concentration on the GDE performance.

The improved TFFA-model presented in this study takes all local electrolyte effects fully into account and, for the first time, includes also a realistic electrolyte distribution based on operando synchrotron experiments. The model is validated with a series of measurements with different oxygen and sodium hydroxide concentrations. Subsequently, the model is used to analyze local transport phenomena within the GDE, the still existing limitations and possible improvement potentials.

4.2 Experimental

The investigated electrodes were produced by means of a wet spraying process which is described in detail in prior publications [4,5]. A dispersion consisting of silver particles (SF9ED, Ferro GmbH), hydroxyethyl methyl cellulose (1 wt.% solution, WALOCEL™ MKX 70000 PP01), a PTFE dispersion (TF 5060GZ, 3M™ Dyneon™) and demineralized water was mixed and applied on a nickel mesh as current collector (106 µm × 118 µm mesh size, 63 µm thickness, Haver & Boecker OHG) with an airbrush gun (Evolution, 0.6 mm pin hole, Harder & Steenbeck). After further hot pressing and sintering, the final electrode was obtained. The examined electrodes in this study consist of 97 wt.% silver and 3 wt.% PTFE as binder. Due to the hand-made production minor deviations between individual electrodes may occur. The electrode for the electrochemical measurements had a catalyst loading, determined by weighing, of 99 $mg_{Ag}\,cm^{-2}$ resulting in a thickness of 250 µm. The corresponding values of the electrode for the electrolyte distribution are 124 $mg_{Ag}\,cm^{-2}$ and 295 µm.

Electrochemical experiments were performed in an in-house half-cell designed for stationary conditions at the GDE [5]. The GDE divided the half-cell in a gas and electrolyte compartment, where the counter electrode, a platinum wire, was placed. Potentials were measured against a reversible hydrogen electrode (RHE, HydroFlex, Gaskatel) directly connected to the GDE through a Luggin capillary. Electrochemical experiments were carried out using a Zennium Pro Potentiostat (Zahner GmbH). Different electrolyte and oxygen concentrations were supplied. The NaOH electrolyte was prepared from caustic flakes (≥ 99 wt.%, Carl Roth) and demineralized water. Electrolyte circulation (32 mL min^{-1}) combined with a large reservoir guaranteed a constant concentration over the entire experiment. The electrolyte concentrations were determined by an automated acid-base titration (G20S, Mettler Toledo) using 0.1 mol L^{-1} hydrochloric acid (Carl Roth). On the gas side dry oxygen diluted by nitrogen in the desired ratio was supplied by mass flow controllers (VDM™, Bronkhorst®) with a total volume flow rate of 100 mL min^{-1}. The oxygen concentration was measured by an OXYMAT 6 analyzer (Siemens) at the gas outlet. Temperatures and pressures of the gas and electrolyte were monitored by sensors and could be controlled individually. All experiments were carried out at 80 °C and 1 bar at the electrolyte and gas side.

Polarization curves were measured with iR-compensation using a steady-state method. In the kinetic region (open circuit potential down to 800 mV vs. RHE), a resolution of 10 mV was chosen, which was increased to 25 mV in the linear region (800 – 200 mV vs. RHE). The steady state was defined as a maximum total deviation of 1 nA s^{-1} or a relative deviation of 0.01 % s^{-1}. To determine the ohmic resistance, pseudo-galvanostatic electrochemical impedance spectroscopy (EIS) was carried out at 1 kA m^{-2} with an amplitude of 5 mV.

In total 15 polarization curves were determined with electrolyte concentrations of 27, 31 and 36.6 wt.% NaOH and oxygen concentrations of 100, 80, 60, 40 and 20 vol.%, respectively. The stability of the electrode was guaranteed by a long term start-up procedure including chronopotentiometric and EIS experiments, until a steady state was reached. After this series of experiments, the measurements with 31 wt.% NaOH electrolyte, which is representative for industrial conditions [2], were reproduced.

4.2.1 Electrochemical operando measurement

To analyze the electrolyte distribution within the GDE independently of the polarization curves experiments, a half-cell compartment was designed and constructed, which allows simultaneous electrochemical characterization and operando X-ray radiography. The construction and application of the half-cell setup is described in detail by Gebhard et al. [6,8] and Paulisch et al. [9]. During the experiments 30 wt.% NaOH electrolyte was pumped in a closed cycle with 33 mL min^{-1} at room temperature while an oxygen gas flow rate of 20 mL min^{-1} was set. The chronoamperometric tests were performed according to the following procedure: 30 min at 0.2 V vs. RHE, 0.9 V vs. RHE, 0.5 V vs. RHE and 0.2 V vs. RHE, respectively.

The experiments were performed at DIAMOND in Oxfordshire at beam line I13-2. The test was carried out with pink beam using a silver filter with a thickness of 0.035 mm to limit the spectrum to energies below the silver K edge and to ensure a high transmission of the sample. For imaging a pcoEdge 5.5 camera with a field of view of 2.1 x 1.8 mm and an effective Pixel size of 0.8 µm, an UPlanFL N objective and a 500 µm LuAG:Ce doped scintillator were applied.

To analyze the images captured during the operando measurements, corresponding calculations were performed using the imaging software Fiji (ImageJ 1.52p) [20,21]. To visualize the electrolyte volume inside the GDE, the radiographic projections were normalized to the reference state of the cell filled with electrolyte, but without an applied potential.

4.3 GDE model

4.3.1 Model structure

The model is divided into four compartments. The gas and the electrolyte compartment are divided by the GDE. The GDE, consisting of silver particles and PTFE forming a complex pore system, is subdivided into a gas transport and reaction layer. The electrolyte can intrude the pore system forming finger-shaped so-called flooded agglomerates where the reaction takes place. A schematic overview of the model is given in Figure 4.1. To describe the processes with a pseudo 2D-model the following assumptions are made. These are based on the original model [15], but important improvements have been made to include the new findings. The assumptions can be summarized as follows.

1. The electrode is operated at steady state at constant temperature and pressure within all regions.
2. On both sides of the electrode, a stagnant diffusion film is assumed, where the mass transport is described by Maxwell-Stefan diffusion. On the gas side the film thickness is z_b, on the electrolyte side z_d.
3. The electrode has an overall thickness of $z_{electrode}$, which is divided into two zones. First, the gas diffusion layer with thickness z_s consisting of gas pores with radius r_s, porosity ε_s and tortuosity τ_s. Second, the reaction layer with thickness z_t consisting of flooded agglomerates and gas pores with radius r_t, porosity ε_t and tortuosity τ_t.
4. The uniformly distributed flooded agglomerates consisting of solid electrode and liquid electrolyte are covered by a liquid thin-film of thickness δ_{tf}. The cylindrically shaped agglomerates have the radius r_{ag}, the porosity ε_n and the tortuosity τ_n.
5. In the gas phase the transport of oxygen, water vapor and nitrogen is described by Maxwell-Stefan and Knudsen diffusion.
6. In the liquid phase, the transport of water, hydroxide and sodium ions is described by Maxwell-Stefan diffusion.
7. Liquid water and vapor are in equilibrium inside the reaction layer. The phase transition is allowed everywhere inside and at the boundaries.
8. In the reaction layer oxygen dissolves in the electrolyte and is in equilibrium at the thin-film surface. The dissolved oxygen diffuses

radially according to Fick's law into the flooded agglomerate, where the electrochemical reaction takes place.

9. The electrochemical reaction is considered as direct four-electron reaction of oxygen and water.

$$O_2 + 2\,H_2O + 4\,e^- \rightarrow 4\,OH^- \qquad (4.1)$$

The possible formation of peroxide intermediate is neglected due to high catalytic decomposition rate on silver [22]. The reaction is described by a simplified Tafel kinetics with a change of the rate determining step at the overpotential η_{change}.

10. The electrochemical reaction is considered to be first order with respect to the oxygen concentration.

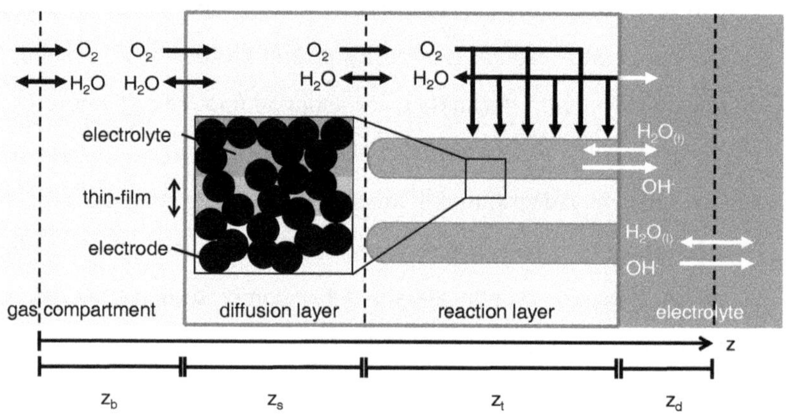

Figure 4.1: Schematic view of the TFFA-model including the main transport fluxes over all domains.

The flooded agglomerates are described as cylindrical fingers with a uniform distribution over the electrode surface (see Figure 4.2). Three of these electrolyte fingers form an equilateral triangle.

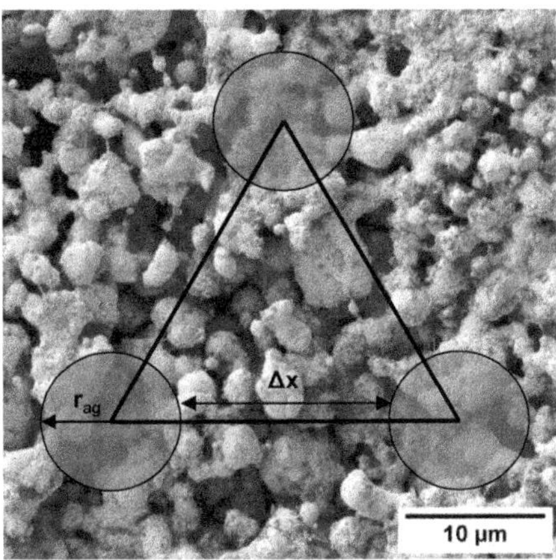

Figure 4.2: Front view of the flooded agglomerates inside the GDE. The uniform distribution can be described by an equilateral triangle. Flooded agglomerates and triangle do correspond to the true dimensions.

Therefore, the whole electrolyte hold-up in the reaction layer can be described by the relative projection surface area of the flooded agglomerates inside the triangle ($S_{\text{electrolyte}}$).

$$S_{\text{electrolyte}} = \frac{2\,\pi\,r_{\text{ag}}^2}{\sqrt{3}\left(2\,r_{\text{ag}} + \Delta x\right)^2} \tag{4.2}$$

As a result, the free pore space inside the reaction layer is given by multiplying the remaining relative surface area with the electrode porosity (eq. 4.3). The porosity of the flooded agglomerates can then be calculated with eq. 4.4.

$$\epsilon_{\text{t}} = \epsilon_{\text{s}}\,(1 - S_{\text{electrolyte}}) \tag{4.3}$$

$$\epsilon_{\text{n}} = \frac{\epsilon_{\text{s}} - \epsilon_{\text{t}}}{1 - \epsilon_{\text{t}}} \tag{4.4}$$

4.3.2 Mass transport

The mass transport in z-dimension is described by a Maxwell-Stefan approach. In the gas phase the driving force is given by the partial pressure gradient of the individual species. The film diffusion in the gas compartment is described by eq. 4.5. Here, *z'* represents the length variable, while the index shows the respective compartment. The Maxwell-Stefan diffusion coefficients are calculated according to Fuller et al. [23], which yields precise results for the given gas composition [24]. Inside the GDE effective diffusion coefficients and overlaying Knudsen diffusion is taken into account. The flux of nitrogen is considered to be zero, as this species does not participate in the reaction. The full set of equations is given in the appendix.

$$-\frac{1}{RT}\frac{dP_i}{dz'_{\mathrm{b}}} = \sum_{\substack{j=1 \\ j\neq i}}^{n} \frac{P_j N_i - P_i N_j}{P_{\mathrm{T,b}} Đ^{\mathrm{g}}_{\mathrm{i,j}}} \tag{4.5}$$

The transport in the liquid phase is also described with the Maxwell-Stefan equation, in this case with the chemical potential as the driving force (eq. 4.6). Similar to nitrogen in the gas phase the flux of sodium ions is considered to be zero, since they do not participate in the reaction. This is only valid for the stationary case. In dynamic processes, the hydroxide ion concentration might change locally, in such a case the charge is balanced by an occurring flux of sodium ions due to electro neutrality condition.

$$-\frac{c_i}{RT}\frac{d\mu_i}{dz'_{\mathrm{t}}} = \sum_{\substack{j=1 \\ j\neq i}}^{n} \frac{c_j N_i - c_i N_j}{c_{\mathrm{T,t}} Đ^{\mathrm{eff}}_{i,j,\mathrm{t}}} \tag{4.6}$$

With the definition of the chemical potential, the driving force can be described using the activity of the components. Together with the electro neutrality condition and the boundary condition of a sodium-ion flux of zero, it is possible to describe the entire liquid transport with eq. 4.7. In contrast to chlor-alkali membrane processes [25] migration as a driving force can be excluded as there are only two charged species that neutralize each other. The activity of water is calculated by the method described by Balej [26]. The required diffusion coefficients are calculated with values given for KOH [27] but with the viscosities of NaOH presented by Olsson et al. [28]. This procedure has already been applied in other studies, as no reliable data for NaOH is available [18]. All electrolyte properties are calculated locally, an overview is given in Table 4.1.

$$-\frac{(1-w_{\mathrm{NaOH}})\rho_{\mathrm{NaOH}}}{M_{\mathrm{H_2O}}}\frac{d\ln\left(a_{\mathrm{H_2O_{(l)}}}\right)}{dz_{\mathrm{t}}'} = \frac{x_{\mathrm{NaOH}}N_{\mathrm{H_2O_{(l)}}}-(1-x_{\mathrm{NaOH}})N_{\mathrm{OH^-}}}{Đ^{\mathrm{eff}}_{\mathrm{OH^-,H_2O,t}}}+\frac{x_{\mathrm{NaOH}}N_{\mathrm{H_2O_{(l)}}}}{Đ^{\mathrm{eff}}_{\mathrm{Na^+,H_2O,t}}} \tag{4.7}$$

The liquid transport in the stagnant film is calculated analogously, the full set of equations can again be found in the appendix. Between the interfaces the fluxes and the electrolyte concentrations are transferred.

4.3.3 Material balance

The total flux of oxygen can be calculated according to Faraday's law with the applied current density. Inside the reaction layer water is present in gaseous and liquid state and can be summarized as the total amount of water. This total water flux is stoichiometrically converted as well as the flux of hydroxide ions (eq. A10, A11). Water is consumed during the reaction as hydroxide ions are generated. The local flux of oxygen is also described by Faraday's law and the local current density in the solid electrode. Beyond the reaction layer, all material fluxes are constant as no other sources or sinks exist. At the interface of the gas transport and reaction layer, the total flux of water converts into gaseous water, as no liquid is assumed to reach the gas transport layer. On the other side at the interface between reaction and liquid transport layer, the total flux of water is present as liquid water, since it is assumed that there is no gaseous species present.

4.3.4 Reaction rates

The electrochemical reaction is described with a classical reaction engineering approach, where the flooded agglomerate is treated similar to a porous catalyst pellet [29]. The catalyst efficiency (η_{eff}) and the Thiele modulus (ϕ, eq. A14) allow to calculate the radial profile of the oxygen concentration. The local reaction rate is described by four equations, which are equally valid in steady state. The local reaction rate is described by the gradient of the oxygen flux (eq. 4.8). On the other hand, the reaction rate is coupled with a local thin-film current density (j_{tf}) through Faraday's law (eq. 4.9). At the interface between gas and electrolyte, the reaction rate is described by the mass transfer of oxygen through the thin-film (eq. 4.10). The equilibrium concentration (c^*_{O2}) is calculated by Henry's law and the required local Henry constant is calculated with the correlations by Tromans [30]. Finally the chemical reaction inside the flooded agglomerate is described by the catalyst pellet approach (eq. 4.11).

$$r = -\frac{dN_{O_2}}{dz'_t} \tag{4.8}$$

$$r = \frac{j_{tf}}{4\,F}\,S_{tf} \tag{4.9}$$

$$r = \frac{D_{L_{O2}}}{\delta_{tf}}\left(c^*_{O_2} - c_{O_2}\right) S_{tf} \tag{4.10}$$

$$r = k_c\,c_{O_2}\,\eta_{eff}\,(1 - \epsilon_t) \tag{4.11}$$

The required specific surface area of the flooded agglomerates (S_{tf}) is obtained from eq. 4.12 for the assumed cylindrical shape of the electrolyte fingers.

$$S_{tf} = \frac{2}{r_{ag}}\,S_{electrolyte}\,\tau_n \tag{4.12}$$

Additionally the specific surface area allows the determination of the surface enhancement factor (S_A), which indicates the increase of the active surface area due to the electrolyte distribution in the porous GDE. This parameter serves as a reference value to other studies.

$$S_A = S_{tf}\,z_t \tag{4.13}$$

The diffusion coefficient of oxygen in the electrolyte (D_{LO2}) depends on the electrolyte concentration. Zhang et al. [31] showed a good agreement between the Stokes-Einstein relation and experimental data at 23 °C for a wide concentration range of the NaOH electrolyte. However, for higher temperatures the viscosity has to be corrected by an exponent as described by Chatenet et al. [32] (eq. 4.14).

$$D_{L_{O_2}} = \frac{k_B T}{6\pi r_{O_2} \eta_{NaOH}^{\alpha}} \tag{4.14}$$

With a radius r_{O_2} of 72 pm and an exponent α of 0.36 the data of Chatenet et al. [32] for the high electrolyte concentration (33.5 wt.% NaOH) and Han et al. [33] for pure water is well reproduced. The chosen radius is much smaller than described by Zhang et al. [31], but in good agreement with reported covalent radii of oxygen [34].

4.3.5 Electrochemical kinetics

The thin-film current density for the reaction rates is described by Tafel equations, in which the rate determining step depends on the local overpotential (eq. 4.15). This phenomenon is well described in several publications on experimental investigations of the ORR kinetics [35–38] as well as in theoretical studies [39].

$$\begin{aligned} i_{\mathrm{tf}} &= A \cdot c_{\mathrm{O_2}} \cdot 10^{\frac{\Delta\varphi}{T_{\mathrm{SA}}}} \qquad \text{for: } \Delta\varphi < \eta_{\mathrm{change}} \\ i_{\mathrm{tf}} &= B \cdot c_{\mathrm{O_2}} \cdot 10^{\frac{\Delta\varphi}{T_{\mathrm{SB}}}} \qquad \text{for: } \Delta\varphi > \eta_{\mathrm{change}} \end{aligned} \tag{4.15}$$

The local overpotential is given by eq. 4.16. The potential of the solid electrode and the electrolyte is calculated independently by Ohm's law (see eq. A17, A18). The temperature and concentration-dependent standard potential (E^0) is calculated by the Nernst equation. The required standard potential under normal conditions (80 °C, pH of 13.996) of the ORR is determined by Bratsch [40], the activities of the species and oxygen concentrations are already given by the mentioned electrolyte properties correlations.

$$\Delta\varphi = E_{\mathrm{electrode}} - E_{\mathrm{electrolyte}} - E^0 \tag{4.16}$$

The Tafel slopes are taken from Kandaswamy et al. [41] and recalculated to the desired temperature of 80 °C assuming a temperature-independent charge transfer coefficient. While the first slope (T_{SA}) was found to have a value of 54 mV, the second slope (T_{SB}) depends on the electrolyte purity and ranges from 113 to 282 mV. The overpotential at the change of the rate determining step (η_{change}) was given with approx. 180 mV, almost independent of the electrolyte purity. The kinetic constant A corresponds to the exchange current density. Since no literature data can be used due to the unknown active area of the GDE, this parameter is therefore fitted to the experimental GDE data. The kinetic constant of the second Tafel equation (*B*) is determined by eq. 4.17, which guarantees continuity and thus numerical stability in the transition section.

$$B = A \cdot 10^{\eta_{\text{change}} \cdot \left(\frac{T_{\text{SB}} - T_{\text{SA}}}{T_{\text{SB}} \cdot T_{\text{SA}}}\right)} \tag{4.17}$$

By linking the electrochemical kinetics with the reaction rate, the local electric and ionic currents are known, and thus the electrode and electrolyte potentials. Finally, the overall overpotential for a given current density is calculated by eq. 4.18.

$$\eta = E_{\text{electrode}}(0) - E_{\text{electrolyte}}(z_t) - E^0(z_t) \tag{4.18}$$

4.3.6 Boundary conditions

At the interfaces between the gas bulk and the stagnant gas film, the composition of the gas feed is introduced as a boundary condition. At the interface between electrolyte bulk and diffusion layer, the electrolyte concentration is known. Furthermore, the applied current density is given and thus the total flux of oxygen. At the interfaces between the individual domains, the gas or electrolyte compositions and fluxes are transferred. As mentioned earlier, the fluxes of nitrogen and sodium ions are zero, as these species do not participate in the reaction. Moreover, it is assumed that there are no liquids in the gas transport layers, and vice versa no gas in the liquid transport layers. In the reaction layer liquid water and vapor are in equilibrium, therefore the partial pressure of water corresponds to the vapor pressure of the electrolyte.

Table 4.1: Overview of locally calculated electrolyte properties and difference to the original TFFA-model [15].

name	symbol	unit	source	previous model [15]
density	ρ	kg m^{-3}	[28]	local
viscosity	μ	Pa s	[28]	for bulk
vapor pressure	p_m	Pa	[28]	local
Henry constant	H	mol m^{-3} Pa^{-1}	[30]	fixed value
activity coefficient of NaOH	γ_{NaOH}	-	[42]	for bulk
activity of water	a_{H_2O}	-	[26]	for bulk
electrolyte conductivity	κ_{NaOH}	S m^{-1}	fitted to [43]	fixed value
Maxwell-Stefan diffusion coefficient	$Đ_{i,j}$	m^2 s^{-1}	[27], KOH data used with NaOH viscosity	for bulk
Oxygen diffusion coefficient in electrolyte	$D_{L_{O2}}$	m^2 s^{-1}	eq. 13 fitted to data from [33] and [32].	fixed value

An overview of all given model parameters and literature data is given in Table 4.2.

Table 4.2: Overview of available model parameters from experimental setup, literature and assumptions.

parameter	value	unit	description	source
z_b	10	mm	thickness of the stagnant gas film in front of the electrode	experimental setup
$z_{electrode}$	250	µm	electrode thickness	measured value
ε_s	0.38		porosity of dry electrode	measured value
τ_s	3.2		tortuosity of dry electrode	assumption
τ_t	3.2		tortuosity of gas channels in reaction zone	assumption
τ_n	3.2		tortuosity of flooded agglomerates	assumption
r_s	0.35	µm	radius of gas channels in gas diffusion layer of the GDE	[5]
r_t	0.35	µm	radius of gas channels in reaction layer of the GDE	[5]
δ_{tf}	20	nm	thin-film thickness	assumption
T_{SA}	54	mV	1st Tafel slope	[41]

4.4 Results and discussion

4.4.1 Operando electrolyte distribution

To determine the electrolyte distribution independently from the polarization curves, the radiography images obtained during operando synchrotron experiments are used. The radiography shows a front view of the electrode after 75 min of operation (Figure 4.3a). The Figure shows a region between the nickel wires of the mesh in a size of 150 x 170 µm. Areas which are colored in yellow and orange represent areas of low absorption, areas in lilac and blue exhibit a high absorption which indicates pores filled with electrolyte. A quantification of the electrolyte volume is not possible due to the application of pink beam. At steady-state electrolyte structures are visible similar to those described in Figure 4.2. Individual flooded agglomerates with a cylindrical shape penetrate the electrode in z-dimension with unknown length. Apart from a few exceptions, there is no merging or interaction with neighboring flooded agglomerates, which substantiates assumption number four. The radii and the distances between the flooded agglomerates were measured using Fiji (ImageJ 1.52p) [20,21] for three different regions. The distances between the individual flooded agglomerates, indicated by the arrows in Figure 4.3a, are distributed around an average value of 13 µm with a slight positive skewness (Figure 4.3b). The average value for the agglomerates radius is approx. 6.5 µm, however, the determination is more challenging as not all flooded agglomerates have an ideal roundness.

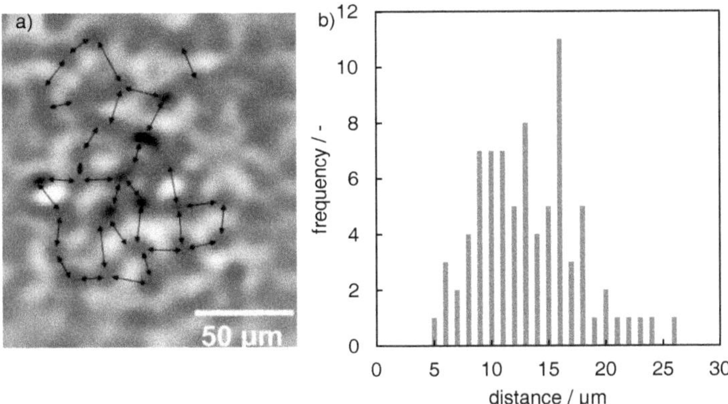

Figure 4.3: a) Example of radiographic stationary electrolyte distribution. Darker spots are filled with electrolyte and are represented in the model as flooded agglomerates. b) Distribution of the measured distances between individual electrolyte spots for a larger sample.

The two model parameters (r_{ag}, Δx), which describe the electrolyte distribution, can be directly derived from the operando experiment. For a clearer visualization a light Gaussian filter was applied, which increases the flooded agglomerates slightly. This results in an error of 1 – 2 pixels, which corresponds to about 1.6 – 3.2 µm, due to the 2-way binning of the images. Therefore, in the model the radius of the flooded agglomerates is fixed to a value of 5 µm, while the distance in between is set to 15 µm. The synchrotron experiments are discussed in more detail in another forthcoming publication by Paulisch et al..

The last unknown parameter for a complete description of the electrolyte distribution is the intrusion depth (z_t). Unfortunately, this depth cannot be determined by the operando experiment. However previous experimental studies showed a droplet formation on the gas side [11] and a constant flow towards the gas compartment [9]. Therefore we assume, that the electrolyte fingers expand over the entire thickness of the electrode. With this intrusion depth of 250 µm a surface enhancement factor (S_A) of 46.4 is obtained, which is comparable to the values (36.4 and 56 respectively) determined by Röhe et al. [17]. The question of the correct choice of the intrusion depth is further discussed in section 4.4.5. An overview of all determined model parameters by the operando experiments is given in Table 4.3.

Table 4.3: Evaluated model parameters by operando experiments.

parameter	value	unit	description	source
z_s	0	µm	length of diffusion layer for gaseous species inside the GDE	assumption
z_t	250	µm	length of reaction layer inside the GDE	assumption
r_{ag}	5	µm	radius of flooded agglomerates	operando
Δx	15	µm	distance between flooded agglomerates	operando
ε_t	0.32		gas porosity in reaction layer	eq. 3
ε_n	0.08		electrolyte porosity in reaction layer	eq. 4
S_{tf}	185733	$m^2\ m^{-3}$	specific surface area of thin-film	eq. 12
S_A	46.43	$m^2\ m^{-2}$	surface enhancement factor	eq. 13

4.4.2 Polarization curves parameterization

For the polarization curves experiments the electrolyte and oxygen concentrations were varied. The best performance was obtained using the 27 wt.% NaOH electrolyte, as ionic conductivity as well as oxygen solubility have the highest values at these conditions. With increasing electrolyte concentration the overvoltages are rising. As expected, lower oxygen concentrations also increase the overpotential. In Figure 4.4 a detailed look on the kinetics is provided while Figure 4.5 shows all polarization curves. For clarity, the reproduced values for 31 wt.% NaOH are not included.

For further parameterization process only 4 model parameters need to be determined, in the kinetic region the kinetic parameter A and the overvoltage for the change in the rate determining step (η_{change}) and for higher current densities the second Tafel slope (T_{SB}). Due to unknown fluid dynamics within the half-cell the thickness of the stagnant film (z_d) is also determined using the polarization curves, taking the respective electrolyte concentration into account. According to fluid dynamics correlations, the stagnant film thickness is proportional to the square root of the fluid's viscosity [44].

The parameterization process starts at the kinetic region for low current densities. Below an overvoltage of 0.3 V the influence of mass transport can be neglected in the model, if the electrolyte distribution is fixed. The first Tafel slope (T_{SA}) is set to a value of 54 mV, recalculated to the desired temperature from the experimental data of Kandaswamy et al. [41]. As the electrolyte distribution is assumed to be known, only the kinetic parameters A, and η_{change} are adjusted using a least squares method for all 15 polarizations curves. For η_{change} a value of 175 mV is determined during the optimization routine, which is in good agreement with the literature values [41]. The parameter A slightly depends on the electrolyte concentration, the corresponding values are summarized in Table 4.4. The simulated and experimentally determined polarization curves show a relatively good agreement for the lower electrolyte concentrations (see Figure 4.4). In the case of the highest electrolyte concentration of 36.6 wt.% larger deviations are identified. Apparently, the value for the first Tafel slope is higher than for the other electrolyte concentrations. Thus, further kinetic measurements are required.

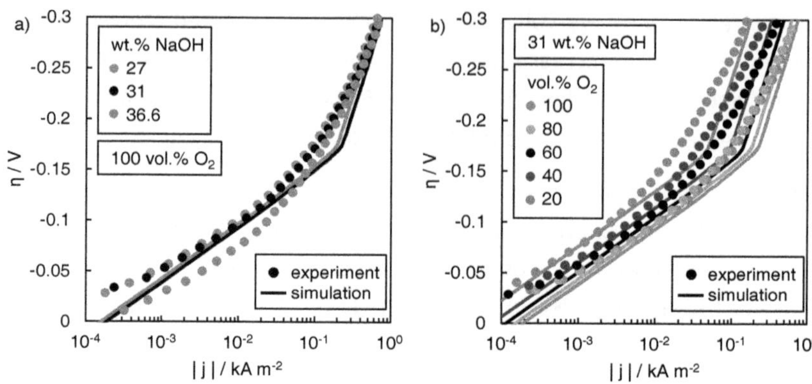

Figure 4.4: Experimentally determined (symbols) and calculated (lines) polarization curves in the kinetic region for 100 vol.% and different electrolyte concentrations (a) and for 31 wt.% NaOH at different oxygen concentrations (b). In diagram (a) the simulated curve for 27 wt.% NaOH is overlaid by the other polarization curves.

For higher current densities the second Tafel slope (T_{SB}) and the thickness of the stagnant electrolyte film (z_d) is adjusted with the same method described above. Finally, a value of 210 mV for the second Tafel slope and stagnant film thicknesses of 225 – 290 µm (see Table 4.4 and 4.5) are determined.

The comparison between measured and calculated polarization curves over the entire range of current densities is shown in Figure 4.5. The measurements at the electrolyte concentrations of 27 and 31 wt.% NaOH are well represented by the model over the whole range of oxygen concentrations. At the highest electrolyte concentration of 36.6 wt.% NaOH, the discrepancy between experiment and model increases again. For all shown oxygen concentrations in Figure 4.5d, the predicted current densities between an overpotential of -0.5 and -0.8 V are too high. However, the limiting current densities are still reliably predicted. One reason for the discrepancy is most probably the limited validity of the correlations describing the electrolyte properties. These correlations had to be extrapolated to higher values, for which no experimental data is available.

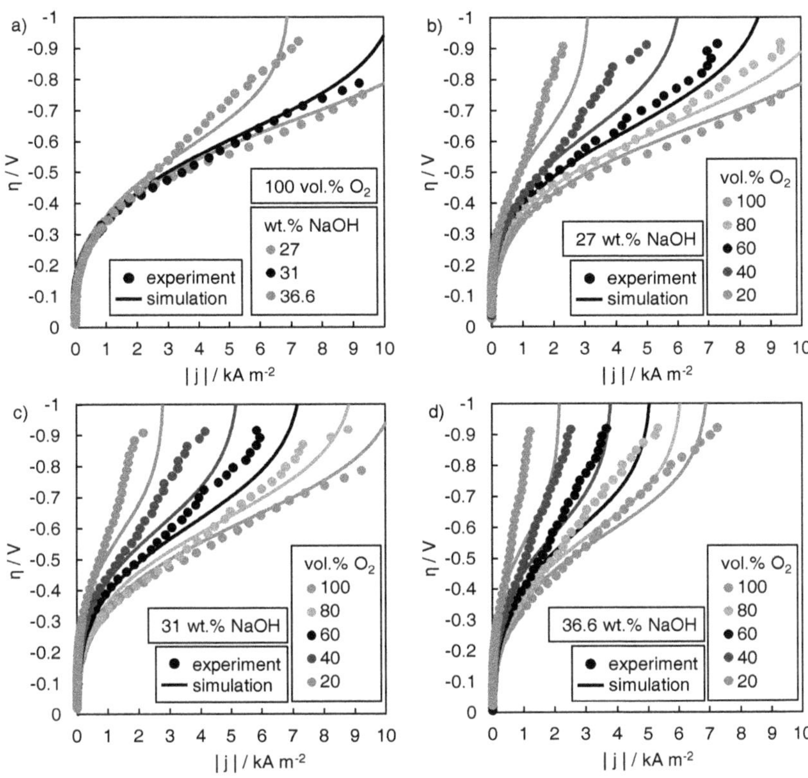

Figure 4.5: Experimentally determined (symbols) and calculated (lines) polarization curves for 100 vol.% and different electrolyte concentrations (a), results at different oxygen concentrations for 27 wt.% (b), 31 wt.% (c) and 36.6 wt.% NaOH (d).

Table 4.4: Overview of model parameters as a function of the electrolyte bulk concentration.

c_{NaOH} / wt.%	A / A m mol^{-1}	z_d / µm
27	0.083	225
31	0.116	250
36.6	0.166	290

Table 4.5: Overview of model parameters derived from electrochemical experiments.

parameter	value	unit	description
T_{SB}	210	mV	Tafel slope for 2nd Tafel equation
η_{change}	175	mV	overvoltage for change in rate determining step

Overall a fair agreement between experimentally determined and simulated polarization curves is reached. However, the deviations rise for higher electrolyte concentrations. This is most probably caused by a lack of reliable kinetic and electrolyte property data. The deviations also increase for lower oxygen concentrations, which may be caused by an underestimation of the gas transport properties or diffusion lengths. Note that the electrolyte distribution was obtained from operando experiments carried out at room temperature and with 30 wt.% NaOH. The influence of temperature and electrolyte concentration on the electrolyte distribution is unknown. Furthermore, the intrusion depth used for the simulations is an assumption, as the value could not be extracted from the operando experiments. Therefore the influence of the electrolyte distribution will be discussed in detail in chapter 4.4.5. Before this discussion, further analysis of the processes inside the GDE will be carried out with the as-determined parameter set.

4.4.3 Model-based analysis of processes inside the GDE

For the following analysis only the industrial base case (31 wt.% NaOH, 100 vol.% O_2) is considered. Figure 4.6a shows the electrolyte concentration over all electrolyte containing domains. With increasing current density the electrolyte concentration rises as more hydroxide ions are produced. In the stagnant electrolyte film in front of the electrode, a linear increase takes place. In the reaction layer there is a strong accumulation of the electrolyte, especially within the first 50 µm of the intrusion depth. The hydroxide ions are produced during the reaction and the transport is hindered by the pore network. At the maximum current density of 10 kA m^{-2} the concentration rises up to 36 wt.% in the stagnant film and to 54 wt.% at the highest intrusion depth, which corresponds to approx. 12 mol L^{-1} and 20 mol L^{-1}, respectively. The calculated values inside the stagnant film are lower than the values measured by Botz et al. [16], who determined electrolyte concentrations in the stagnant film above 17 mol L^{-1}. However, in their experimental setup no convection and thus, no removal of the hydroxide ions from the electrode surface took place. Consequently, the stagnant film in the experiments by Botz et al. [16] is supposed to be thicker, allowing a higher accumulation in front of the electrode. Taking these differences into account, we can conclude that our simulations are in good agreement with experimental findings.

Looking at the local reaction rate (Figure 4.6b) the most active part of the electrode is directly at the interface towards the electrolyte. While at lower current densities the whole flooded agglomerate is active, the most reactive part shifts towards the electrolyte with rising current density. The high electrolyte concentrations inside the GDE lead to various effects. First of all, the vapor pressure of the electrolyte, and thus the partial pressure of gaseous water, decreases, which leads to an increase of the partial pressure of oxygen. The original version of the model [15] therefore predicted a performance increase with higher electrolyte concentration contrary to the new experimental results. In the improved model presented here, liquid and oxygen diffusion coefficients, electrolyte conductivity and oxygen solubility decrease at higher electrolyte concentrations, leading to a significant decrease of the electrode performance. As the electrolyte concentration inside the GDE increases due to higher current densities, the most reactive zone shifts towards the electrolyte.

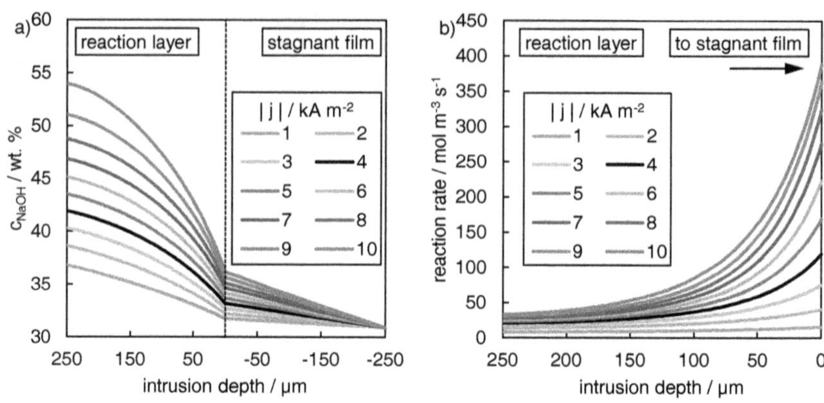

Figure 4.6: a) Electrolyte concentration as a function of the intrusion depth. Positive values of the depth denote the region inside the GDE, negative values represent the stagnant electrolyte film. b) Reaction rate as a function of the intrusion depth for different current densities.

4.4.4 Analysis of molar fluxes

The model also allows a deeper look into details of the molar fluxes inside the electrode. In Figure 4.7 the main fluxes inside the flooded part of the electrode are shown for a typical industrial operation at 31 wt.% NaOH, 100 vol.% oxygen, and a current density of 4 kA m^{-2}. Oxygen enters the reaction layer and the flux decreases according to the local reaction rate reaching zero at the interface of the reaction layer and the stagnant film. The total water flux also decreases due to the stoichiometric conversion, while the hydroxide ion flux evolves. However, a small flux of water still remains in the direction of the gas compartment. In the shown case this flux results in the formation of a condensate amount of 1.17 g day^{-1} cm^{-2}, which is in good agreement with experimental values reported in earlier studies [4]. Equally interesting are the fluxes of liquid and gaseous water. Directly at the interface of the reaction layer and stagnant film a huge amount of water evaporates and a flux of gaseous water towards the gas compartment is created. On the other hand the flux of liquid water faces towards the electrolyte bulk almost through the whole reaction layer and is reversed directly at the interfaces towards the stagnant film. This means, that the flux of liquid water helps to remove the reaction product of hydroxide ions from the reaction layer as both species flow in the same direction, thus boosting the GDE performance. The efficient removal of hydroxide ions leads a higher utilization of the flooded agglomerate. To create this supportive liquid water flux the water evaporates at the interface to stagnant film, is transported towards the gas compartment and condensates on the electrolyte finger where the flow direction is reversed towards the electrolyte again. This water circle shows the complexity of the transport mechanisms inside the reaction layer and that at least two model dimensions are needed to describe the simultaneous flux of liquid and gaseous water. We believe that this is the reason why the intrusion depth of the electrolyte assumed in our model is much higher than predicted by Röhe et al. for their dynamic simulations with a one-dimensional model [17].

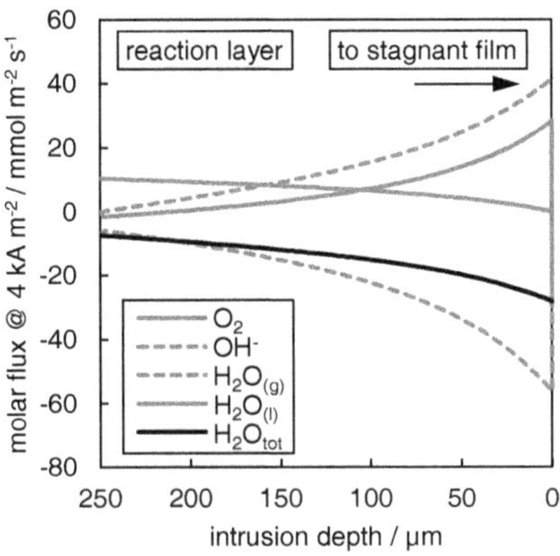

Figure 4.7: Molar fluxes in the electrode as a function of the intrusion depth. Positive fluxes direct towards the electrolyte, negative fluxes towards the gas compartment.

4.4.5 Electrolyte distribution studies

Although the operando experiments show an almost uniform distribution of flooded agglomerates over the whole electrode, the real electrolyte distribution is still associated with a high uncertainty. The real intrusion depth cannot be measured by this technique and electrowetting effects [45] are neglected. Moreover, the operando measurements were carried out at room temperature and the electrochemical measurements at 80 °C. Therefore we investigate in this section the influence of the parameters defining the electrolyte distribution (z_t, r_{ag}, Δx) on the GDE performance. Due to the complex interaction between these parameters only the individual contribution of one parameter is examined.

The greatest uncertainty lies in the assumption that the electrolyte intrudes over the entire thickness of the electrode. A reduction of the reaction layer thickness leads to higher overvoltages in the model (Figure 4.8a). Reducing the intrusion depth to 125 µm has only a minor effect and decreases the limiting current density to 9.8 kA m^{-2}. At further reduced intrusion depth, the overvoltages increase strongly. A reduction of the distance between the flooded agglomerates, and thus an increase of the number of the agglomerates and the corresponding surface enhancement factor, boosts the GDE performance significantly (Figure 4.8b). Without any distance between the individual flooded agglomerates, but still with good oxygen supply in the small remaining gas channels, almost no signs of mass transport limitation are visible and the electrode performance would be primarily defined by the electrochemical kinetics. Reducing the radius of the flooded agglomerates also decreases the electrode performance (see Figure 4.8 c). Although the oxygen transport into smaller agglomerates is more efficient, the GDE performance is mainly determined by the reduced available specific surface area between gas and electrolyte (eq. 4.2), at least in the range of radii depicted in Figure 4.8c. Even larger radii would invert the trend and deteriorate the GDE performance (not shown).

For all the presented cases the overvoltage at a current density of 4 kA m^{-2} is taken and plotted in Figure 4.8d as a function of the calculated surface enhancement factor (S_A) (eq. 4.13). Obviously, there is no clear correlation between the surface enhancement factor and the resulting overvoltage. A change in the electrolyte distribution not only changes the specific surface area, but also available spaces for gas and electrolyte transport, which both are responsible for the already mentioned water circulation boosting the electrode performance. This is not considered if the three-phase interface is reduced to only one single parameter.

According to the present study, it is likely that the electrolyte expands over a considerable part of the GDE. Moreover, a higher intrusion depth leads to an enlargement of the three-phase boundary and enhances the electrode performance. It remains therefore unclear why the electrodes with a barrier layer presented by Gebhard et al. [6] exhibited a better performance than uniformly composed GDE. It could be that this barrier layer leads to a more finely distributed electrolyte in the accessible part of the GDE, or that the water circulation mechanism is enhanced which accelerates the removal of hydroxide ions. To be able to give a clear interpretation, further investigations with these model electrodes are required.

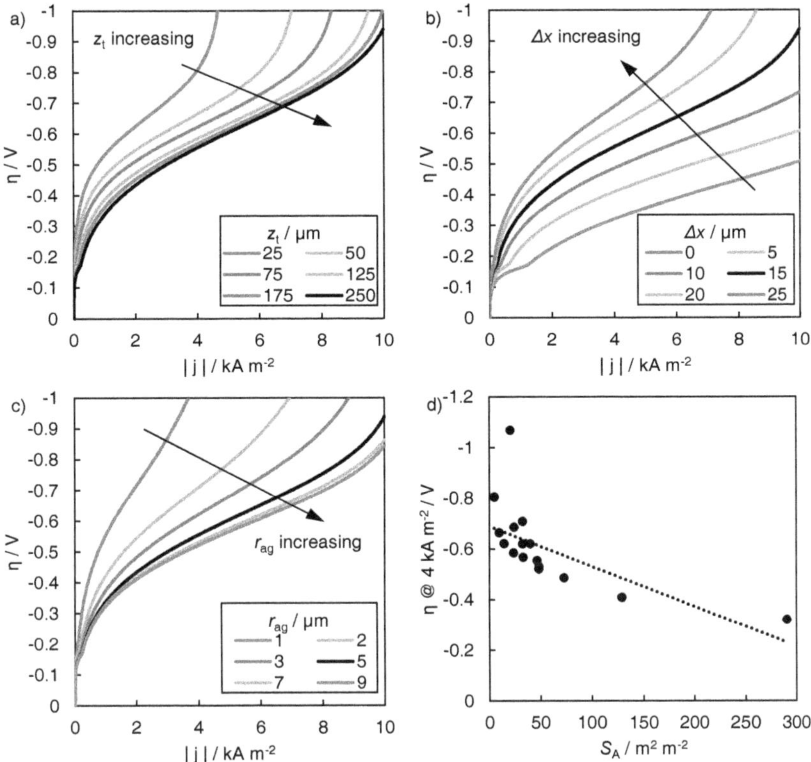

Figure 4.8: Influence of the electrolyte distribution on the polarization curve for the industrial base case (31 wt.% NaOH, 100 vol.% O_2). Intrusion depth (a), distance between flooded agglomerates (b), radius of flooded agglomerates (c) and the resulting overvoltage at industrial current density as a function of the resulting surface enhancement factors (d).

The above parameter study shows, that the electrolyte distribution is most important for the model. Therefore, more detailed measurements of the electrolyte distribution are crucial for model validation. Not only the process conditions, but also the electrode potential (electrowetting) will very likely have an effect on the distribution. It could be also the case that a part of the electrode is completely flooded as assumed by Röhe et al. [17,18]. This would increase the diffusion length of the liquid electrolyte and reduce the length of the reaction layer.

4.5 Conclusion

The presented pseudo-2D TFFA-model allows a detailed look inside GDEs during ORR in highly alkaline electrolyte. For the first time a realistic electrolyte distribution, based on operando synchrotron measurements, could be implemented. For the model validation, steady state polarization curves obtained at a temperature of 80 °C, different NaOH concentrations (27 – 36.6 wt.%) and oxygen contents (20 – 100 vol.%) were used. Through adjustment of only two remaining parameters - the exchange current density, which was found to be dependent on the electrolyte concentration, and the thickness of the stagnant film in the electrolyte compartment - a quite good agreement between measurements and model predictions was achieved. The model-based analysis of the processes inside the GDE revealed that the intrusion depth of the electrolyte in the GDE is a further open question. Until now, the model predicts that the electrolyte penetrates between half and the full thickness of the electrode. This question should be tackled with additional operando investigations, but also through measurements with model electrodes having a defined thickness of the silver electrocatalyst, which is intruded by the electrolyte. Results obtained with these model electrodes will be presented in a forthcoming publication.

The model also shows that the simultaneously available transport channels for gas and liquid species are very important for the performance of the GDE. According to our analysis, evaporated water inside the GDE condensates again on the flooded agglomerates. The water cycle driven by this process strongly enhances the removal of the hydroxide ions from the flooded parts of the GDE and boosts the electrode performance. This is why the intrusion depth of the electrolyte in our model is much higher than predicted by one-dimensional GDE models. We strongly believe that the complexity of the transport mechanisms inside GDE requires least two dimensions for proper description. The final goal would be a 2D or even 3D model coupling a realistic picture of the pore network with the electrochemical reaction. To achieve this level of modeling, however, the electrolyte distribution must first be determined even more precisely.

4.6 Acknowledgements

Financial support for this study by Deutsche Forschungsgemeinschaft in the framework of the research unit "Multiscale analysis of complex three-phase systems: Oxygen reduction at gas-diffusion electrodes in aqueous electrolyte"(FOR 2397; TU 89/13-1, MA 5039/3-1). The Authors also thank Christoph Rau and Shashidhara Marathe for the support at the synchrotron beam line I13-2 as well as André Hilger and Markus Osenberg for their support in the evaluation process of the synchrotron data.

4.7 Appendix

For the gas transport inside the GDE Knudsen diffusion is also taken into account and effective diffusion coefficients are considered.

$$-\frac{1}{RT}\frac{dP_i}{dz'_{\mathrm{s}}}=\frac{N_i}{Đ^{\mathrm{eff}}_{i,\mathrm{K,s}}}+\sum_{\substack{j=1\\ j\neq i}}^{n}\frac{P_jN_i-P_iN_j}{P_{\mathrm{T,s}}Đ^{\mathrm{eff}}_{i,j,\mathrm{s}}} \tag{A1}$$

Analogously, gas transport is described in the reaction layer.

$$-\frac{1}{RT}\frac{dP_i}{dz'_{\mathrm{t}}}=\frac{N_i}{Đ^{\mathrm{eff}}_{i,\mathrm{K,t}}}+\sum_{\substack{j=1\\ j\neq i}}^{n}\frac{P_jN_i-P_iN_j}{P_{\mathrm{T,t}}Đ^{\mathrm{eff}}_{i,j,\mathrm{t}}} \tag{A2}$$

The transport in the liquid stagnant film in front of the electrode is described using the chemical potential as the driving force.

$$\begin{aligned}-\frac{(1-w_{\mathrm{NaOH}})\rho_{\mathrm{NaOH}}}{M_{\mathrm{H_2O}}}\frac{d\ln\left(a_{\mathrm{H_2O_{(l)}}}\right)}{dz'_{\mathrm{d}}}\\ =\frac{x_{\mathrm{NaOH}}N_{\mathrm{H_2O_{(l)}}}-(1-x_{\mathrm{NaOH}})N_{\mathrm{OH^-}}}{Đ_{\mathrm{OH^-,H_2O}}}+\frac{x_{\mathrm{NaOH}}N_{\mathrm{H_2O_{(l)}}}}{Đ_{\mathrm{Na^+,H_2O}}}\end{aligned} \tag{A3}$$

Knudsen diffusion coefficient are calculated using the kinetic gas theory.

$$D_{i,\mathrm{K},x} = \frac{2}{3} r_x \sqrt{\frac{8RT}{\pi M_i}} \qquad \text{with } x = \mathrm{s,t} \tag{A4}$$

All effective diffusion coefficients are calculated using the corresponding porosity and tortuosity.

$$D_x^{\mathrm{eff}} = \frac{\epsilon_x}{\tau_x} D_x \qquad \text{with } x = \mathrm{s,t,n} \tag{A5}$$

The oxygen flux is calculated using Faraday's Law and the applied current density.

$$N_{\mathrm{O_2}}(0) = \frac{j}{4F} \tag{A6}$$

The flux of total water in the reaction layer is composed of gaseous and liquid water.

$$N_{\mathrm{H_2O}} = N_{\mathrm{H_2O_{(l)}}} + N_{\mathrm{H_2O_{(g)}}} \tag{A7}$$

The hydroxide ions produced during the reaction and the water consumed are in stoichiometric proportion to the oxygen flux in the reaction layer.

$$N_{\mathrm{OH^-}}(z_\mathrm{t}) = 4N_{\mathrm{O_2}}(0) \tag{A8}$$

$$N_{\mathrm{H_2O}}(0) - N_{\mathrm{H_2O}}(z_\mathrm{t}) = 2N_{\mathrm{O_2}}(0) \tag{A9}$$

This results in the following material balances.

$$N_{\mathrm{OH^-}}(z) = N_{\mathrm{OH^-}}(z_t) - 4N_{\mathrm{O_2}}(z) \tag{A10}$$

$$N_{\mathrm{H_2O}}(z) = N_{\mathrm{H_2O}}(z_t) + 2N_{\mathrm{O_2}}(z) \tag{A11}$$

The local electric current density (j_e) can be calculated using Faraday's law and the local flux of oxygen.

$$j_\mathrm{e}(z) = 4FN_{\mathrm{O_2}}(z) \tag{A12}$$

The equilibrium concentration of oxygen ($c^*_{\mathrm{O_2}}$) is calculated by Henry's law.

$$c^*_{\mathrm{O_2}}(z) = H(z)P_{\mathrm{O_2}}(z) \tag{A13}$$

Thiele modulus and thus the catalyst efficiency for eq. 4.11 is calculated from the ratio of reaction and diffusion rate. These equations are valid for planar geometries, but can be applied for cylindrical shapes as well, if half of the radius is used as the characteristic length.

$$\phi = \frac{r_{\mathrm{ag}}}{2} \sqrt{\left(\frac{k_{\mathrm{c}}}{D_{\mathrm{L_{O_2}}}^{\mathrm{eff}}}\right)} \tag{A14}$$

$$\eta_{\mathrm{eff}} = \frac{1}{\phi} \qquad \text{for: } \phi \geq 3 \tag{A15}$$

$$\eta_{\mathrm{eff}} = \frac{\tanh(\phi)}{\phi} \qquad \text{for: } \phi < 3 \tag{A16}$$

The potentials of the solid electrode and electrolyte are calculated according to Ohm's law. For the efficient conductivity of the electrolyte porosity and tortuosity of the flooded agglomerates are considered.

$$\frac{dE_{\mathrm{electrode}}}{dz_{\mathrm{t}}'} = -\frac{j_{\mathrm{e}}}{\kappa_{\mathrm{e}}} \tag{A17}$$

$$\frac{dE_{\mathrm{electrolyte}}}{dz_{\mathrm{t}}'} = -\frac{j_{\mathrm{i}}}{\kappa_{\mathrm{NaOH}}^{\mathrm{eff}}} \tag{A18}$$

$$\kappa_{\mathrm{NaOH}}^{\mathrm{eff}} = \frac{\epsilon_{\mathrm{t}}}{\tau_{\mathrm{t}}} \kappa_{\mathrm{NaOH}} \tag{A19}$$

The sum of electrical current density of the solid electrode (j_{e}) and ionic current density of the electrolyte (j_{i}) corresponds at every location to the applied current density.

$$j = j_{\mathrm{e}} + j_{\mathrm{i}} \tag{A20}$$

4.8 References

[1] J. Kintrup, M. Millaruelo, V. Trieu, A. Bulan, E.S. Mojica, Gas Diffusion Electrodes for Efficient Manufacturing of Chlorine and Other Chemicals, Electrochem. Soc. Interface 26 (2017) 73–76. https://doi.org/10.1149/2.F07172if.

[2] Chlor-Alkali Electrolyis, https://ucpcdn.thyssenkrupp.com/_legacy/UCPthyssenkruppBAISUhdeChlorineEngineers/assets.files/products/chlor_alkali_electrolysis/thyssenkrupp_chlor_alkali_brochure_web.pdf, accessed 22 October 2020.

[3] N. Furuya, H. Aikawa, Comparative study of oxygen cathodes loaded with Ag and Pt catalysts in chlor-alkali membrane cells, Electrochim. Acta 45 (2000) 4251–4256. https://doi.org/10.1016/S0013-4686(00)00557-0.

[4] I. Moussallem, S. Pinnow, N. Wagner, T. Turek, Development of high-performance silver-based gas-diffusion electrodes for chlor-alkali electrolysis with oxygen depolarized cathodes, Chem. Eng. Process. 52 (2012) 125–131. https://doi.org/10.1016/j.cep.2011.11.003.

[5] D. Franzen, B. Ellendorff, M.C. Paulisch, A. Hilger, M. Osenberg, I. Manke, T. Turek, Influence of binder content in silver-based gas diffusion electrodes on pore system and electrochemical performance, J. Appl. Electrochem. 49 (2019) 705–713. https://doi.org/10.1007/s10800-019-01311-4.

[6] M. Gebhard, T. Tichter, D. Franzen, M.C. Paulisch, K. Schutjajew, T. Turek, I. Manke, C. Roth, Improvement of Oxygen-Depolarized Cathodes in Highly Alkaline Media by Electrospinning of Poly(vinylidene fluoride) Barrier Layers, ChemElectroChem 7 (2020) 830–837. https://doi.org/10.1002/celc.201902115.

[7] P. Frania, Herstellung, Analyse und Optimierung von Sauerstoffverzehrkathoden mit elektrochemisch abgeschiedenem Silberkatalysator zum Einsatz in der Chlor-Alkali-Elektrolyse. Dissertation, 1st ed., 2016.

[8] M. Gebhard, M. Paulisch, A. Hilger, D. Franzen, B. Ellendorff, T. Turek, I. Manke, C. Roth, Design of an In-Operando Cell for X-Ray and Neutron Imaging of Oxygen-Depolarized Cathodes in Chlor-Alkali Electrolysis, Materials 12 (2019). https://doi.org/10.3390/ma12081275.

[9] M.C. Paulisch, M. Gebhard, D. Franzen, A. Hilger, M. Osenberg, N. Kardjilov, B. Ellendorff, T. Turek, C. Roth, I. Manke, Operando Laboratory X-Ray Imaging of Silver-Based Gas Diffusion Electrodes during Oxygen Reduction Reaction in Highly Alkaline Media, Materials 12 (2019). https://doi.org/10.3390/ma12172686.

[10] P. Kunz, M. Paulisch, M. Osenberg, B. Bischof, I. Manke, U. Nieken, Prediction of Electrolyte Distribution in Technical Gas Diffusion Electrodes: From Imaging to SPH Simulations, Transp. Porous Med. 132 (2020) 381–403. https://doi.org/10.1007/s11242-020-01396-y.

[11] P. Jeanty, C. Scherer, E. Magori, K. Wiesner-Fleischer, O. Hinrichsen, M. Fleischer, Upscaling and continuous operation of electrochemical CO_2 to CO conversion in aqueous solutions on silver gas diffusion electrodes, J. CO2 Util. 24 (2018) 454–462. https://doi.org/10.1016/j.jcou.2018.01.011.

[12] M.B. Cutlip, An approximate model for mass transfer with reaction inporous gas diffusion electrodes, Electrochim. Acta 20 (1975) 767–773. https://doi.org/10.1016/0013-4686(75)85013-4.

[13] X.-L. Wang, S. Koda, Scale-up and Modeling of Oxygen Diffusion Electrodes for Chlorine-Alkali Electrolysis I. Analysis of Hydrostatic Force Balance and Its Effect on Electrode Performance, Denki Kagaku 65 (1997) 1002–1013. https://doi.org/10.5796/kogyobutsurikagaku.65.1002.

[14] X.-L. Wang, S. Koda, Scale-up and Modeling of Oxygen Diffusion Electrodes for Chlorine-Alkali Electrolysis II. Effects of the Structural Parameters on the Electrode Performance Based on the Thin-Film and Flooded-Agglomerate Model, Denki Kagaku 65 (1997) 1014–1025. https://doi.org/10.5796/kogyobutsurikagaku.65.1014.

[15] S. Pinnow, N. Chavan, T. Turek, Thin-film flooded agglomerate model for silver-based oxygen depolarized cathodes, J. Appl. Electrochem. 41 (2011) 1053–1064. https://doi.org/10.1007/s10800-011-0311-2.

[16] A. Botz, J. Clausmeyer, D. Öhl, T. Tarnev, D. Franzen, T. Turek, W. Schuhmann, Local Activities of Hydroxide and Water Determine the Operation of Silver-Based Oxygen Depolarized Cathodes, Angew. Chem. Int. Ed. 57 (2018) 12285–12289. https://doi.org/10.1002/anie.201807798.

[17] M. Röhe, A. Botz, D. Franzen, F. Kubannek, B. Ellendorff, D. Öhl, W. Schuhmann, T. Turek, U. Krewer, The Key Role of Water Activity for the Operating Behavior and Dynamics of Oxygen Depolarized Cathodes, ChemElectroChem 6 (2019) 5671–5681. https://doi.org/10.1002/celc.201901224.

[18] M. Röhe, F. Kubannek, U. Krewer, Processes and Their Limitations in Oxygen Depolarized Cathodes: A Dynamic Model-Based Analysis, ChemSusChem 12 (2019) 2373–2384. https://doi.org/10.1002/cssc.201900312.

[19] N.S. Vasile, R. Doherty, A.H.A. Monteverde Videla, S. Specchia, 3D multi-physics modeling of a gas diffusion electrode for oxygen reduction reaction for electrochemical energy conversion in PEM fuel cells, Appl. Energy 175 (2016) 435–450. https://doi.org/10.1016/j.apenergy.2016.04.030.

[20] D. Legland, I. Arganda-Carreras, P. Andrey, MorphoLibJ: integrated library and plugins for mathematical morphology with ImageJ, Bioinformatics 32 (2016) 3532–3534. https://doi.org/10.1093/bioinformatics/btw413.

[21] J. Schindelin, I. Arganda-Carreras, E. Frise, V. Kaynig, M. Longair, T. Pietzsch, S. Preibisch, C. Rueden, S. Saalfeld, B. Schmid, J.-Y. Tinevez, D.J. White, V. Hartenstein, K. Eliceiri, P. Tomancak, A. Cardona, Fiji: an open-source platform for biological-image analysis, Nat. Methods 9 (2012) 676–682. https://doi.org/10.1038/nmeth.2019.

[22] P.K. Adanuvor, R.E. White, Oxygen Reduction on Silver in 6.5M Caustic Soda Solution, J. Electrochem. Soc. 135 (1988) 2509–2517. https://doi.org/10.1149/1.2095367.

[23] E.N. Fuller, K. Ensley, J.C. Giddings, Diffusion of halogenated hydrocarbons in helium. The effect of structure on collision cross sections, J. Phys. Chem. 73 (1969) 3679–3685. https://doi.org/10.1021/j100845a020.

[24] B.E. Poling, J.M. Prausnitz, J.P. O'Connell (Eds.), The properties of gases and liquids, 5th ed., McGraw-Hill, New York, NY, 2001.

[25] R.R. Sijabat, M.T. de Groot, S. Moshtarikhah, J. van der Schaaf, Maxwell–Stefan model of multicomponent ion transport inside a monolayer Nafion membrane for intensified chlor-alkali electrolysis, J. Appl. Electrochem. 49 (2019) 353–368. https://doi.org/10.1007/s10800-018-01283-x.

[26] J. Balej, Water vapour partial pressures and water activities in potassium and sodium hydroxide solutions over wide concentration and temperature ranges, Int. J. Hydrog. Energy 10 (1985) 233–243. https://doi.org/10.1016/0360-3199(85)90093-X.

[27] J. Newman, D. Bennion, C.W. Tobias, Mass Transfer in Concentrated Binary Electrolytes, Berich. Bunsen. Gesell. 69 (1965) 608–612. https://doi.org/10.1002/bbpc.19650690712.

[28] J. Olsson, Å. Jernqvist, G. Aly, Thermophysical properties of aqueous NaOH-H_2O solutions at high concentrations, Int. J. Thermophys. 18 (1997) 779–793. https://doi.org/10.1007/BF02575133.

[29] P. Björnbom, Modelling of a double-layered PTFE-bonded oxygen electrode, Electrochim. Acta 32 (1987) 115–119. https://doi.org/10.1016/0013-4686(87)87018-4.

[30] D. Tromans, Oxygen solubility modeling in inorganic solutions: concentration, temperature and pressure effects, Hydrometallurgy 50 (1998) 279–296. https://doi.org/10.1016/S0304-386X(98)00060-7.

[31] C. Zhang, F.-R.F. Fan, A.J. Bard, Electrochemistry of oxygen in concentrated NaOH solutions: solubility, diffusion coefficients, and superoxide formation, J. Am. Chem. Soc. 131 (2009) 177–181. https://doi.org/10.1021/ja8064254.

[32] M. Chatenet, M. Aurousseau, R. Durand, Comparative Methods for Gas Diffusivity and Solubility Determination in Extreme Media: Application to Molecular Oxygen in an Industrial Chlorine–Soda Electrolyte, Ind. Eng. Chem. Res. 39 (2000) 3083–3089. https://doi.org/10.1021/ie000044g.

[33] P. Han, D.M. Bartels, Temperature Dependence of Oxygen Diffusion in H_2O and D_2O, J. Phys. Chem. 100 (1996) 5597–5602. https://doi.org/10.1021/jp952903y.

[34] B. Cordero, V. Gómez, A.E. Platero-Prats, M. Revés, J. Echeverría, E. Cremades, F. Barragán, S. Alvarez, Covalent radii revisited, Dalton Trans. (2008) 2832–2838. https://doi.org/10.1039/b801115j.

[35] B.B. Blizanac, P.N. Ross, N.M. Markovic, Oxygen electroreduction on Ag(111): The pH effect, Electrochim. Acta 52 (2007) 2264–2271. https://doi.org/10.1016/j.electacta.2006.06.047.

[36] B.B. Blizanac, P.N. Ross, N.M. Marković, Oxygen reduction on silver low-index single-crystal surfaces in alkaline solution: rotating ring disk$_{(Ag(hkl))}$ studies, J. Phys. Chem. B 110 (2006) 4735–4741. https://doi.org/10.1021/jp056050d.

[37] M.D. Obradović, B.N. Grgur, S.L. Gojković, L.M. Vračar, Enhancement of the electrochemical reduction of oxygen at platinum by nickel underpotential deposition, J. Solid State Electrochem. 11 (2006) 77–83. https://doi.org/10.1007/s10008-005-0072-0.

[38] D.B. Sepa, M.V. Vojnovic, L.M. Vracar, A. Damjanovic, Apparent enthalpies of activation of electrodic oxygen reduction at platinum in different current density regions—II. Alkaline solution, Electrochim. Acta 31 (1986) 97–101. https://doi.org/10.1016/0013-4686(86)80068-8.

[39] T. Shinagawa, A.T. Garcia-Esparza, K. Takanabe, Insight on Tafel slopes from a microkinetic analysis of aqueous electrocatalysis for energy conversion, Sci. Rep. 5 (2015) 13801. https://doi.org/10.1038/srep13801.

[40] S.G. Bratsch, Standard Electrode Potentials and Temperature Coefficients in Water at 298.15 K, J. Phys. Chem. Ref. Data 18 (1989) 1–21. https://doi.org/10.1063/1.555839.

[41] S. Kandaswamy, A. Sorrentino, S. Borate, L.A. Živković, M. Petkovska, T. Vidaković-Koch, Oxygen reduction reaction on silver electrodes under strong alkaline conditions, Electrochim. Acta 320 (2019) 134517. https://doi.org/10.1016/j.electacta.2019.07.028.

[42] J. Balej, Activity Coefficients of Aqueous Solutions of NaOH and KOH in Wide Concentration and Temperature Ranges, Collect. Czech. Chem. C. 61 (1996) 1549–1562. https://doi.org/10.1135/cccc19961549.

[43] T.F. O'Brien, T.V. Bommaraju, F. Hine, Handbook of chlor-alkali technology, Springer, New York, 2005.

[44] H. Oertel (Ed.), Prandtl - Führer durch die Strömungslehre, Springer Fachmedien Wiesbaden, Wiesbaden, 2012.

[45] F. Mugele, J.-C. Baret, Electrowetting: from basics to applications, J. Phys. Condens. Matter 17 (2005) R705-R774. https://doi.org/10.1088/0953-8984/17/28/R01.

5 Experimental and model-based analysis of electrolyte intrusion depth in silver-based gas-diffusion electrodes

Reproduced and adapted under the creative common (CC BY 4.0) license.

D. Franzen, C. Krause, T. Turek;
ChemElectroChem, 8, 2186–2192 (2021)
https://doi.org/10.1002/celc.202100278

Table of Content

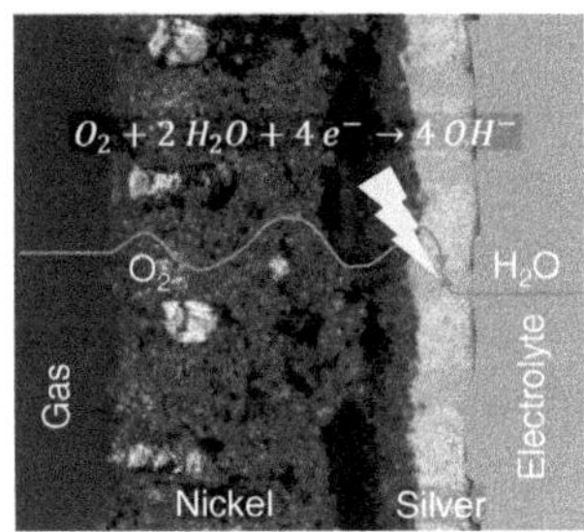

Finding the reaction zone: Silver-based gas-diffusion electrodes (GDE) are generally used for the oxygen reduction reaction in highly alkaline media. Despite wide research interest and successful utilization of these electrodes on industrial scale, the exact location of the reaction remains uncertain. By employing model electrodes with defined reaction zone, it is possible to gain a better insight into the processes inside GDE.

Abstract

The electrolyte distribution is a central point of discussion for understanding the processes inside gas-diffusion electrodes (GDE) for the oxygen reduction reaction in highly alkaline media. During first radiographic operando synchrotron experiments, the liquid electrolyte was located, however, the through-plane distribution remains unclear. Therefore, model electrodes consisting of nickel and silver layers are developed to determine the electrolyte intrusion depth. Nickel-based GDE are modified to achieve a pore system morphology suitable for the oxygen reduction reaction and subsequently coated with silver-PTFE catalyst layers. These graded electrodes form gas-diffusion (nickel) and reaction (silver) layers. The electrodes performance is determined under industrial conditions (80 °C, 30 wt.% NaOH electrolyte) as a function of the silver layer thickness and thus of the effective intrusion depth of the electrolyte. The model-based analysis confirms the experimental determined intrusion depths. Nevertheless, additional operando tomography measurements would help to further improve the understanding of the processes inside GDE.

Keywords: gas-diffusion electrode, oxygen reduction reaction, graded electrode, modelling, electrolyte distribution

5.1 Introduction

Chlorine is one of the most important basic materials for the chemical industry, which is mainly produced by the very energy intensive chlor-alkali electrolysis [1]. By introducing the so-called oxygen depolarized cathode (ODC) to the state-of-the art membrane process, electrical energy savings of up to 25 % are obtained [2]. The ODC consumes oxygen and suppresses the hydrogen evolution reaction, which results in a reduction of the required cell voltage of up to 1 V. Due to the low solubility of oxygen at operation conditions (80 °C, 30-32 wt.% NaOH) [3], the ODC is designed as a gas-diffusion electrode (GDE). The GDE consist of an electrocatalyst and a hydrophobic binder, allowing the electrolyte to intrude the electrode, but provide enough hydrophobicity to prevent a complete flooding of the electrode. In this way, a complex electrolyte distribution inside the GDE is formed, determining the three-phase boundary and thus the available surface area for the electrochemical reaction. Therefore, precise knowledge of the electrolyte distribution within GDE is fundamental. In industrial ODCs, silver is utilized as the electrocatalyst, as the oxygen reduction reaction (ORR) activity on silver is comparable to platinum at the given process conditions [4]. Operando experiments are extremely challenging, as the high silver content leads to a high absorption of X-ray or synchrotron radiation. Until now only radiographic images of the electrolyte distribution inside an ODC during operation are available [5–7]. Unfortunately, these measurements do not provide sufficient information about the penetration depth of the electrolyte.

To further improve the understanding of processes inside GDE, we developed a so-called thin-film flooded agglomerate (TFFA) model, described in detail in our previous publication [7]. This model, developed on the basis of the work of Cutlip [8], Wang and Koda [9,10] and Pinnow et al. [11], describes the ORR in a porous silver-based GDE for the first time with a realistic electrolyte distribution based on operando experiments. However, the penetration depth of the electrolyte remains unknown. In experimental studies, a droplet formation on the gas side of the electrode is observed [5,6,12], and identified to contain electrolyte [13] and not only condensed water vapor. These findings indicate, that the electrolyte intrudes over the full thickness of the electrode. Furthermore, Frania [14] showed with his model electrodes, consisting of silver dendrites in polytetrafluoroethylene (PTFE) membranes, that the silver catalyst is utilized within the entire electrode. However, these membrane-supported electrodes have a completely different structure than conventional GDE made from silver and PTFE particles. While Kunz et al. [15] hardly found any electrolyte imbibition in a silver-based pore system without application of a large differential pressure in their simulations, Röhe et al. [16,17] determined small intrusion depths of the electrolyte based on their dynamic model approach. In summary, quite contradictory information can be found in the literature, although the electrolyte distribution and the penetration depth is of fundamental importance for understanding the processes in GDE with liquid electrolyte. As long as no reliable operando tomography data is available, other indirect methods must be employed.

In a recent study Gebhard et al. [18] tried to prevent a too deep electrolyte intrusion by barrier layers inside the GDE. However, the reported performance boost contradicts our model-based analysis, according to which a larger penetration depth should enhance the electrode performance, and is presumably due to percolation processes inside the special electrode with barrier layer. In a new approach, we therefore manufactured graded electrodes consisting of two layers made of different materials. Nickel-PTFE, which is catalytically almost inactive towards the ORR [19], is used as additional diffusion layer, which is coated with a standard silver-PTFE layer. During operation the electrolyte will likely spread through the graded electrodes with a distribution similar to a standard full silver-PTFE GDE. However, the electrochemical reaction will only take place at the catalytically active silver part. In this way we are able to investigate the effect of the catalyst layer thickness, and thus the effective intrusion depth of the electrolyte, experimentally and can compare the results with our TFFA-model approach.

5.2 Results and Discussion

5.2.1 Analysis of the pore system

In a previous publication, we already presented the importance of an appropriate pore system for the performance of GDEs [20]. Therefore, we tried to prepare nickel and silver layers with comparable pore systems. Koj et al. [21] reported mean pore diameters of approximately 2 µm and a wide distribution of up to 10 µm for pure nickel GDEs. For the oxygen reduction reaction (ORR) in silver-based electrodes, pore diameters of 0.7 µm are best suited. In order to obtain similar pore networks for the nickel and silver layers, the recipe for the nickel suspension is modified as described in the experimental section. Additionally, an intermediate pressing step decreases the pore size of the pure nickel electrodes to a mean value of 1 µm with largest pores having sizes smaller than 2 µm, a reduction of roughly 8 µm. In this way, nickel layers, close to the properties of the silver layer can be produced. The pore size distributions for all electrodes obtained by capillary flow porometry are shown in Figure 5.1. With loadings above 30 mg_{Ag} cm^{-2} the pore system is primarily defined by the silver layer. Mean pore sizes are similar to the pure silver GDE (136 mg_{Ag} cm^{-2}), only the shape of the distribution is slightly influenced by the nickel layer (see results for the pure nickel GDE). The distribution of the pure nickel electrode and the electrode with a loading of 15 mg_{Ag} cm^{-2} is almost identical. We assume that the silver layer is not completely dense, leading to preferred pathways and channeling effects through the catalyst layer. Thus the pore size distribution is primarily defined by the nickel layer. However, for the electrochemical characterization these channeling effects are secondary, as there is no differential pressure between electrolyte and gas compartment during the experiments. Overall, we assume that the electrolyte distribution in the individual electrode areas, nickel and silver layers, does not deviate significantly from each other. Thus, the thickness of the silver layer should define the catalytically active area for the ORR. This allows us to drawn direct conclusions about the effective penetration depth of silver electrodes.

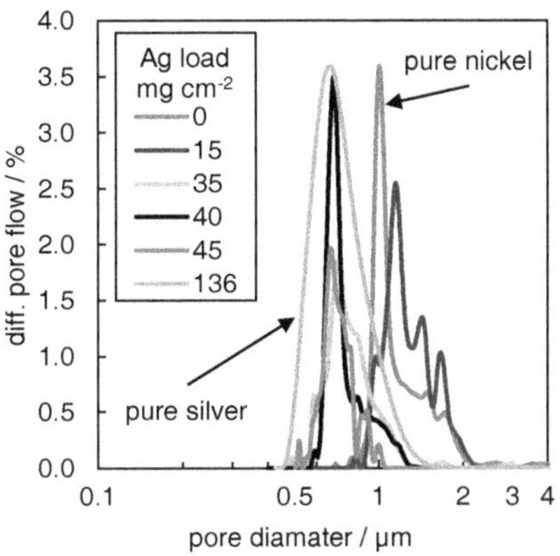

Figure 5.1: Pore size distribution for GDE with different silver loadings.

5.2.2 Layer thickness analysis

The thickness of the silver layer is determined by a nine-point measurement with a thickness gauge after the intermediate pressing step and final sintering of the electrode. This should yield reliable results since the sintering process hardly influences the total thickness of the electrode. In a second approach, thicknesses of the silver layer are calculated with the applied silver loading and the experimentally determined true density of pure silver GDE [20], assuming a linear correlation between silver loading and layer thickness. The values calculated in this way are lower than those measured with the thickness gauge (Figure 5.2). However, the determination of the silver loading is influenced by the average mass loss of the GDE during the sintering process, as the silver layers are applied on wet nickel electrodes. An exact determination of the dry silver load by weighing is therefore not possible due to the production procedure. Furthermore, the influence of the contact line between nickel and silver layers on the true density of the silver layer is unknown. Such effects cannot be ruled out, as deviations between expected and true density were already observed in our previous studies [20]. In a third approach, the layer thicknesses are determined by cross-sectional images at different positions of the electrode (see Figure 5.2 and also Figure 5.6 in the experimental section). Except for the highest silver loading of 45 mg cm^{-2}, the values from thickness gauge and cross-section images are in quite good agreement. However, the error bars for the imaging method indicate the relatively large uncertainties of this method. Moreover, especially at higher loadings, a deformation of the silver layer during the cutting process might take place. Taking all factors into account, the measurements with the thickness gauge should be the most precise. Overall, the goal to prepare silver layers with different thickness on top of a nickel layer with comparable pore structure could be successfully reached.

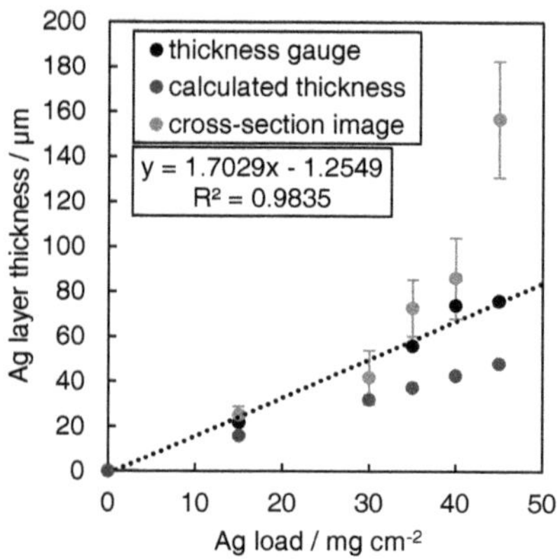

Figure 5.2: Silver layer thickness as a function of the silver loading determined by different evaluation methods.

5.3 ORR activity of graded electrodes

The iR-compensated polarization curves for all investigated electrodes are shown in Figure 5.3. As expected, a pure nickel electrode shows a very poor performance towards the ORR in both, the linear and the kinetic region. The limiting current density is only 0.5 kA m^{-2} and in the kinetic region, the overpotential is approximately 200 mV higher than for all silver-containing electrodes. Already the lowest silver loading of 15 mg_{Ag} cm^{-2} shifts the overpotential in the kinetic region down to 100 mV, showing the improved catalytic activity of the electrode. However, the performance is still limited by the available three-phase boundary as the limiting current density rises to only 1.2 kA m^{-2}. Doubling the loading to 30 mg_{Ag} cm^{-2} increases the current density further to 2.7 kA m^{-2}. Even higher loadings lead to a strong reduction of the overpotentials and to much higher limiting current densities. All electrodes with loadings above 35 mg_{Ag} cm^{-2} are capable to provide an industrially relevant current density of 4 kA m^{-2}. The electrode with a loading of 45 mg_{Ag} cm^{-2} exhibits overpotentials very similar to the pure silver electrode. The only difference is the limiting current density, which is reached at 7.9 kA m^{-2}, while the pure silver electrode does not show any limitations within the investigated range of up to 10 kA m^{-2}. In the kinetic region, a similar effect is observed. For the GDEs with a loading of 15, 30 and 45 mg_{Ag} cm^{-2}, a gradual decrease of the overvoltage by approximately 25 mV is observed. The GDEs with loadings of 35 and 40 mg_{Ag} cm^{-2}, as well as the pure silver electrode, show slightly higher overvoltages. Overall, the trend here seems to be similar to that in the linear region, higher catalyst loadings will lead to lower overpotentials. The inconsistency here is likely to be explained by the hand spraying process. The reproducibility of the electrodes properties is very high, but not perfect. However, this affects the kinetic region more than the linear range and limiting current densities. Further, smaller deviations between the curves of the pure silver and the graded electrodes may occur, as the pore system of the nickel layer is comparable to, but not identical with that of the silver layers. Therefore, there may be slight differences in the electrolyte distribution, which will be neglected in the model-based analysis.

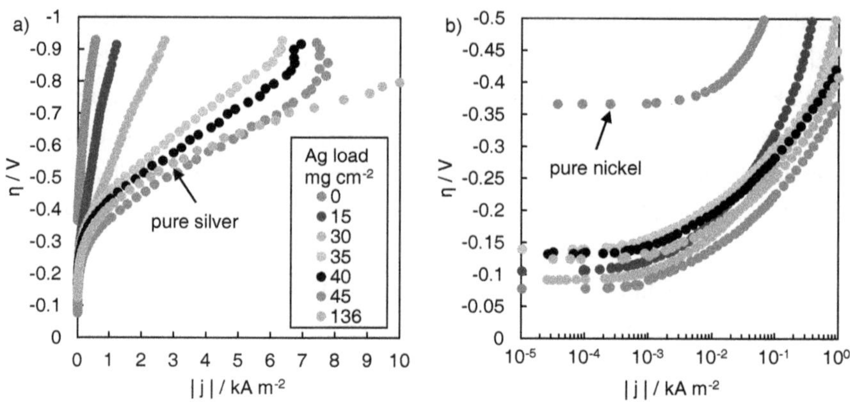

Figure 5.3: Polarization curves for different silver loadings in linear (a) and semi-logarithmic (b) representation.

5.3.1 Model-based analysis

In a first attempt, the silver layer thicknesses determined by the gauge are implemented in the model as the length of the reaction zone, which is intruded by the electrolyte. In reality, as mentioned above, the electrolyte will intrude the nickel layer as well. However, the nickel surface does not contribute to the electrochemical reaction, thus the length of the reaction zone corresponds to the effective electrolyte intrusion depth. In principle the length of the diffusion layer could influence the oxygen transport and thus the oxygen availability at the reaction zone. However, longer diffusion lengths in the gas phase barely influence the overall polarization characteristics, as the transport in the gas phase is much faster than in the liquid phase. Therefore, we assume that the electrodes performance is primarily defined by the effective intrusion depth and that other effects are negligible. The full nickel electrode is neglected in this analysis. The resulting polarization curves are shown in Figure 5.4. For the full silver electrode (136 mg_{Ag} cm^{-2}) two cases, intrusion of the electrolyte over the full and over half the electrode thickness, are considered. A fair agreement between experiment and simulation is reached concerning the linear region and limiting current density, the assumption of a full intrusion represents the experiments slightly better at this point. In the kinetic region, a rather large deviation is evident assuming a full intrusion. The gap between experiment and simulation reduces with decreased intrusion depth of the electrolyte. However, it has to be noted that the model was not parameterized to the exact half-cell conditions and electrolyte concentration used in the experiment. We can therefore conclude that the intrusion depth of the electrolyte in a full silver electrode is most likely between half and full thickness of the GDE, as already proposed in our previous publication [7]. For the higher loadings above 35 $mg_{Ag}\ cm^{-2}$ the trend of the simulation matches the experiment quite well. However, the predicted limiting current densities are slightly higher than during the experiments. For the loadings of 15 and 30 $mg_{Ag}\ cm^{-2}$, the simulation does not fit the experimental data with the predicted limiting current density being up to three times higher than in the experiment. This indicates either that there is another resistance in the GDE system, which is not yet included in the model, or that the available reaction zone is smaller than assumed. In the kinetic region a qualitative agreement between model and experiment is reached. The exact values are not met, but the presumed trend of lower overpotentials with increased silver loading is correctly represented, especially for loadings of 15, 30 and 45 $mg_{Ag}\ cm^{-2}$, respectively. Additionally, the applied Tafel kinetics are invalid at overpotentials above approx. -50 mV indicated by the grey shaded area in the semi-logarithmic representation. Deviations between simulation and experiments in this range are to be expected.

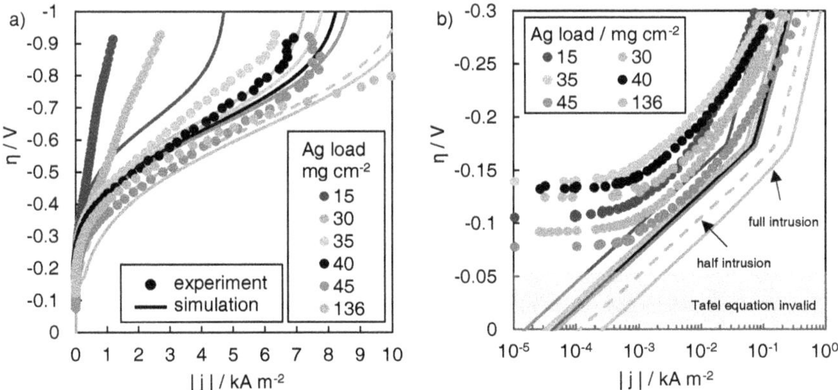

Figure 5.4: Experimentally determined and simulated polarization curves for graded and full silver electrodes in linear (a) and semi-logarithmic (b) representation.

In the second attempt, the length of the reaction layer, i.e. the effective intrusion depth of the electrolyte, is used as a fitting parameter for each electrode and determined via a least squares fit. All other parameters and process variables are kept constant. In this case the full silver electrode is neglected. In Figure 5.5 the determined reaction lengths and resulting polarization curves compared to the experimental data are shown. Except for a silver loading of 30 $mg_{Ag}\ cm^{-2}$ the model-based analysis is in good agreement with the calculations utilizing the density and the thickness gauge measurements. For loadings above 35 $mg_{Ag}\ cm^{-2}$ the maximum deviation to the results of one of these methods is only 5 µm, whereas it rises up to 10 µm for a loading of 15 $mg_{Ag}\ cm^{-2}$. As the only outlier remains a loading of 30 $mg_{Ag}\ cm^{-2}$ where the difference almost reaches 20 µm. The resulting polarization curves now show a fair agreement with the experimental data. In all cases, the limiting current densities are predicted correctly and for catalyst loadings of 35 and 40 $mg_{Ag}\ cm^{-2}$, even a quantitative match is obtained.

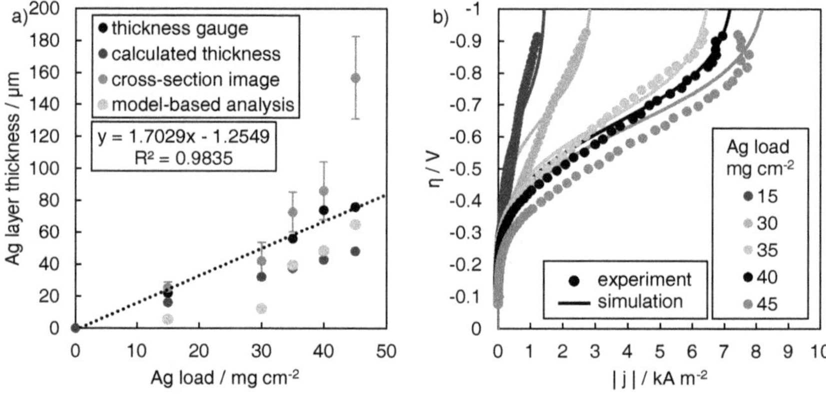

Figure 5.5: Silver layer thickness determined by experiments and model-based analysis (a), and resulting polarization curves (b) as a function of the silver loading.

These experiments combined with the model-based analysis allow a prediction of the active area of GDE. However, there are still some uncertainties regarding the employed parameters in the model. As the half-cell experiments are carried out in the commercially available half-cell, the stagnant electrolyte film in front of the electrode may be thicker than in the flow cell utilized in our previous publication [7]. This influences the diffusion lengths of the liquid species and thus the limiting current density in the model. Further, we assumed that the gas transport in the nickel structure is not inhibited by a possible presence of electrolyte in the pore system. However, considering such effects would increase the number of model parameters as well as the uncertainties and thus would require a new parameterization process of the system. Overall, the experiments and the model-based analysis reveal the key importance of the electrolyte distribution inside the electrodes and how the overall performance is influenced.

5.4 Conclusion

In this study, we developed graded electrodes to determine the catalytically active area of GDEs during ORR experimentally. In contrast to former approaches using barrier layers [18], the division into nickel and silver layers allows an exact prediction of the active part of the electrodes. In an optimized manufacturing process, nickel transport layers were coated with silver catalyst of uniform thickness. We determined the silver layer thickness by a thickness gauge, weighing and verified it by analyzing cross-section images. These thicknesses were implemented in our TFFA-model as the length of the reaction layer. In this model-based analysis, we obtained similar effective penetration depths of the electrolyte. However, due to the manufacturing process we are limited to a maximum catalyst layer thickness of 75 µm in order to ensure the equivalence of the individual electrodes. Experimental data for higher reaction lengths in graded electrodes is not available. Even if a fabrication of catalyst layers thicker than 75 µm was possible, the resulting polarization curves would merge, making it difficult to differentiate between single electrodes. Comparing the graded electrode with the highest silver loading (45 mg_{Ag} cm^{-2}) and the full silver electrode, there is still a significant difference in the limiting current density. This proves, that the active part of the full silver electrode is thicker than 75 µm and thus the electrolyte must intrude much deeper than expected from other simulation approaches [15-17]. Unfortunately, the electrolyte penetration depth cannot be determined exactly for the full silver electrode and will only be revealed by operando tomography experiments. Nevertheless, we proved experimentally and with our model-based analysis that large parts of the full silver GDE are intruded by the electrolyte and are utilized for the ORR in high alkaline media.

5.5 Experimental Section

5.5.1 Electrode preparation

The electrodes are prepared batch wise by means of a wet spraying process inspired by the work of Moussallem et al. [13] and described in detail in previous publications [20,21]. A metal containing suspension is sprayed on a nickel mesh (106 µm x 118 µm mesh size, 63 µm wire thickness, Haver & Boecker OHG) as support and conducting agent using an airbrush gun (Evolution, 0.6 mm pin hole, Harder & Steenbeck). A heating table allows to simultaneously dry each layer while being applied resulting in a homogeneous electrode surface. The spraying process is completed when a total metal loading of approx. 130 mg_{Me} cm^{-2} is reached. Mechanical stability is increased by subsequent hot pressing (LaboPress P200S, Vogt, 15MPa, 130 °C) and sintering in an air furnace (15 min, 330 °C), during which the final pore system is formed.

For the nickel layer a suspension consisting of 30 g Nickel powder (Ni 99.9 %, APS 3 – 7 µm, Alfa Aesar), 30 g hydroxyethyl methyl cellulose (1 wt.% solution, WALOCEL™ MKX 70000 PP01) as thickener and pore building agent, a PTFE dispersion (TF 5060GZ, 3M™ Dyneon™) and 40 g demineralized water is mixed. The sprayed electrodes are hot pressed (58 MPa, 130 °C) in an intermediate step to ensure a smooth surface for the silver coating. This ensures a uniform thickness of the silver layers. For the catalytically active layer a suspension consisting of 30 g silver flakes (SF9ED, Ames Advanced Materials Corp.), 50 g hydroxyethyl methyl cellulose, PTFE dispersion and 40 g demineralized water is mixed. The suspension is applied on one side of the previously prepared nickel electrodes until the desired silver loading is obtained. Afterwards the described processing steps are performed and the final electrode is reached. All electrodes layers, nickel and silver, consist of 97 wt.% metal and 3 wt.% PTFE. For comparison, a pure nickel and a pure silver electrode are also produced.

5.5.2 Physical characterization

The median thickness of the silver layers are determined from nine measurements using a thickness dial gauge (FD 50, Käfer GmbH) before the coating and after final sintering. Additionally, the layer thickness is calculated using the silver loading and the true density of pure silver GDE determined in a former study [20]. Loadings are determined by weighing, taking the average weight loss during the sintering process into account. After electrochemical experiments, layer thicknesses are verified through cross-section images (see Figure 5.6) using a 3D microscope (VHS-2000D, Keyence). Before cutting, the electrodes were frozen using liquid nitrogen to prevent the silver layers from deformation. Flow-through pores are determined using capillary-flow porometry (Porometer 3G, Quantachrome). On top of the probe a wetting fluid (Porofil, Quantachrome) is applied. Inside the device, the fluid is driven out of the pores using a pressure gradient while the nitrogen flow rate is detected. A comparison of nitrogen flow through the wet and dry probe provides the pore size distribution.

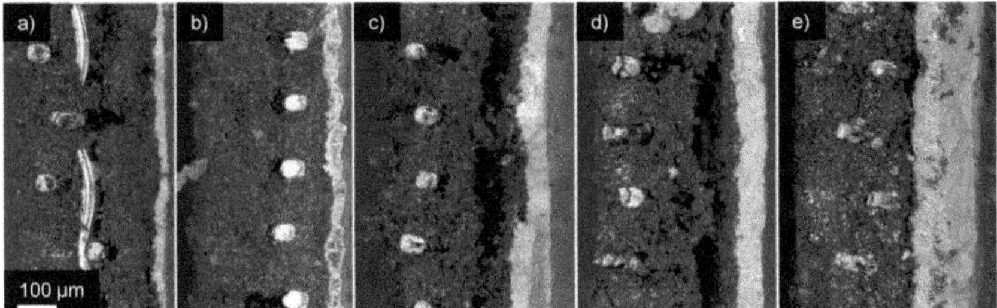

Figure 5.6: Cross-section images of graded nickel/silver electrodes with 15 (a), 30 (b), 35 (c), 40 (d) and 45 mg_{Ag} cm^{-2} (e) loading. The bright layers are silver, the dark areas nickel. In the center, the nickel mesh support is visible.

5.5.3 Electrochemical characterization

Electrochemical experiments are performed in a commercially available half-cell (FlexCell HZ-PP01, Gaskatel GmbH) at 80 °C utilizing an effective surface area of 3.14 cm^2. As electrolyte 30 wt. NaOH solution prepared from caustic flakes (≥99 wt%, Carl Roth) and demineralized water is used. Dry oxygen is supplied with a flow rate of 50 mL min^{-1} and backpressure using a 1 mm water column at the outlet. Potentials are measured using a three-electrode configuration, with the GDE as working, a platinum wire as counter and a reversible hydrogen electrode (RHE), connected through a Luggin-capillary, as reference electrode with a Reference 3000™ potentiostat (Gamry Instruments). The standard potential of the oxygen reduction reaction (ORR) vs. RHE at the given conditions is 1.13 V, calculated by the method introduced in a previous publication [20] The dependence of the potential on temperature [22], activity of water [23] as well as hydroxide ions [24] and the actual oxygen concentration in the electrolyte [3] is considered in the corresponding Nernst equation.

In a start-up procedure, constant current densities (1, 2, 3 and 4 kA m^{-2}) are applied for 5 minutes each, followed by pseudo-galvanostatic impedance spectroscopy (1 kA m^{-2}, 5 mV amplitude) (not shown). The electrode performance is finally investigated by iR-compensated (current interrupt method) linear sweep voltammetry (LSV) measurements starting at open cell potential (OCP) to 200 mV vs. RHE with a scan rate of 0.5 mV s^{-1}.

5.5.4 TFFA-model description

The TFFA-model used to analyze the experiments is described in detail in our previous publication [7]. Maxwell-Stefan diffusion equations are used to describe the gas and liquid transport. The required diffusion coefficients are calculated according to Fuller et al. [25] for the gas phase and as described in Newman et al. [26] for the liquid phase. It is assumed that the liquid electrolyte penetrates the GDE forming a finger-shaped distribution creating the reaction layer. Gas and liquid transport take place simultaneously in the reaction layer. Electrolyte properties, such as vapor pressure [23], Henry constant [3], activity of NaOH [24] and water [23] are spatially resolved. Oxygen can dissolve anywhere on the electrolyte finger. The reaction is described according to Tafel kinetics [27] coupled with a classical reaction engineering approach allowing to calculate the concentration profile of oxygen inside the electrolyte finger [28].

No changes regarding the parameter set are made, and only the boundary conditions are adapted to the experimental setup. In the first attempt, the determined silver layer thickness of each electrode is implemented as the reaction layer length and total intrusion depth of the electrolyte. In the second attempt, the length of the reaction layer is treated as a fitting parameter to achieve the best possible description of the measured results. A schematic overview of the model is shown in Figure 5.7. The complete model code, including all equations and parameters, is provided in the supporting information.

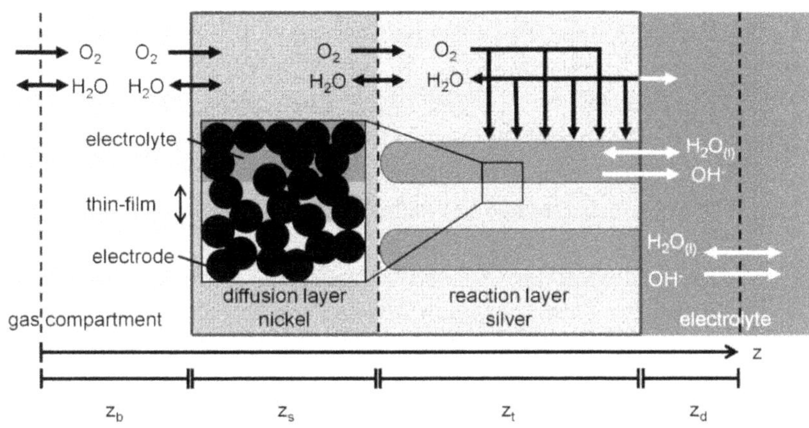

Figure 5.7: Scheme of the used TFFA-model.

5.6 Acknowledgements

The authors acknowledge financial support for this study by Deutsche Forschungsgemeinschaft in the framework of the research unit "Multiscale analysis of complex three-phase systems: Oxygen reduction at gas-diffusion electrodes in aqueous electrolyte" (FOR 2397; research Grant TU 89/13-1) and the open access funding enabled and organized by Projekt DEAL.

5.7 References

[1] J. Kintrup, M. Millaruelo, V. Trieu, A. Bulan, E. S. Mojica, *Electrochem. Soc. Interface* **2017**, *26*, 73.

[2] "Chlor-Alkali Electrolyis", https://ucpcdn.thyssenkrupp.com/_legacy/UCPthyssenkruppBAISUhdeChlorineEngineers/assets.files/products/chlor_alkali_electrolysis/thyssenkrupp_chlor_alkali_brochure_web.pdf.

[3] D. Tromans, *Hydrometallurgy* **1998**, *50*, 279.

[4] N. Furuya, H. Aikawa, *Electrochim. Acta* **2000**, *45*, 4251.

[5] M. Gebhard, M. Paulisch, A. Hilger, D. Franzen, B. Ellendorff, T. Turek, I. Manke, C. Roth, *Materials* **2019**, *12*.

[6] M. C. Paulisch, M. Gebhard, D. Franzen, A. Hilger, M. Osenberg, N. Kardjilov, B. Ellendorff, T. Turek, C. Roth, I. Manke, *Materials* **2019**, *12*.

[7] D. Franzen, M. C. Paulisch, B. Ellendorff, I. Manke, T. Turek, *Electrochim. Acta* **2021**, 137976.

[8] M. B. Cutlip, *Electrochim. Acta* **1975**, *20*, 767.

[9] X.-L. Wang, S. Koda, *Denki Kagaku* **1997**, *65*, 1002.

[10] X.-L. Wang, S. Koda, *Denki Kagaku* **1997**, *65*, 1014.

[11] S. Pinnow, N. Chavan, T. Turek, *J. Appl. Electrochem.* **2011**, *41*, 1053.

[12] P. Jeanty, C. Scherer, E. Magori, K. Wiesner-Fleischer, O. Hinrichsen, M. Fleischer, *J. CO_2 Util.* **2018**, *24*, 454.

[13] I. Moussallem, S. Pinnow, N. Wagner, T. Turek, *Chem. Eng. Process.* **2012**, *52*, 125.

[14] P. Frania, *Herstellung, Analyse und Optimierung von Sauerstoffverzehrkathoden mit elektrochemisch abgeschiedenem Silberkatalysator zum Einsatz in der Chlor-Alkali-Elektrolyse*. Dissertation, 1st ed., Verlag Dr. Hut, **2016**.

[15] P. Kunz, M. Paulisch, M. Osenberg, B. Bischof, I. Manke, U. Nieken, *Transp. Porous Med.* **2020**, *132*, 381.

[16] M. Röhe, F. Kubannek, U. Krewer, *ChemSusChem* **2019**, *12*, 2373.

[17] M. Röhe, A. Botz, D. Franzen, F. Kubannek, B. Ellendorff, D. Öhl, W. Schuhmann, T. Turek, U. Krewer, *ChemElectroChem* **2019**, *6*, 5671.

[18] M. Gebhard, T. Tichter, D. Franzen, M. C. Paulisch, K. Schutjajew, T. Turek, I. Manke, C. Roth, *ChemElectroChem* **2020**, *7*, 830.

[19] S. N.S. Goubert-Renaudin, A. Wieckowski, *J. Electroanal. Chem.* **2011**, *652*, 44.

[20] D. Franzen, B. Ellendorff, M. C. Paulisch, A. Hilger, M. Osenberg, I. Manke, T. Turek, *J. Appl. Electrochem.* **2019**, *49*, 705.

[21] M. Koj, J. Qian, T. Turek, *Int. J. Hydrog. Energy* **2019**, *44*, 29862.

[22] S. G. Bratsch, *J. Phys. Chem. Ref. Data* **1989**, *18*, 1.

[23] J. Balej, *Int. J. Hydrog. Energy* **1985**, *10*, 233.

[24] J. Balej, *Collect. Czech. Chem. C.* **1996**, *61*, 1549.

[25] E. N. Fuller, K. Ensley, J. C. Giddings, *J. Phys. Chem.* **1969**, *73*, 3679.

[26] J. Newman, D. Bennion, C. W. Tobias, *Berich. Bunsen. Gesell.* **1965**, *69*, 608.

[27] S. Kandaswamy, A. Sorrentino, S. Borate, L. A. Živković, M. Petkovska, T. Vidaković-Koch, *Electrochim. Acta* **2019**, *320*, 134517.

[28] P. Björnbom, *Electrochim. Acta* **1987**, *32*, 115.

6 Concluding discussion

In the present doctoral thesis, the overpotential during ORR in silver-based GDE is investigated systematically. Existing model approaches describing the reaction and transport processes inside the GDE were extended and improved. The three selected peer-reviewed manuscripts of the previous chapters thereby describe the approach to this achievement.

Analysis of the pore space in silver-based GDE

In chapter 3 the pore space inside the investigated GDEs is examined in detail. Different electrodes were prepared with a systematic variation of the PTFE binder content. These electrodes were analyzed on the one hand with imaging methods (FIB/SEM) and on the other hand with simple integral methods (porometry). The FIB/SEM revealed a complex pore network with multiple connections inside the GDE. Remarkable is the fact, that in all electrodes, independent of the PTFE content, a similar silver skeleton is formed, taking up approx. 60 % of the available volume. The remaining 40 % are divided into open pore space and deposited PTFE, depending on the total PTFE content of the electrode. Utilizing the capillary-flow porometry the pore size distribution of the nearly pure silver skeleton (99 wt.% Ag, 1 wt.% PTFE) and the stepwise clogging obtained by increasing the overall PTFE content is revealed. This demonstrates a simple and fast method to characterize the pore system of GDEs. The same characterization method was employed to optimize the pore system of the nickel GDE as described in chapter 5.

The electrochemical experiments revealed, that a certain amount of PTFE is necessary in order to ensure a proper electrochemical operation. Electrodes containing of only 1 wt.% PTFE showed a poor electrochemical performance, probably due to the flooding with electrolyte. The oxygen cannot reach the catalytic active sides. Electrodes consisting of 9 and 10 wt.% PTFE also showed a poor performance. We assumed, that the electrolyte cannot intrude the electrode reliably and therefore hardly any reaction sites are formed. Electrodes with PTFE contents between 2 and 7 wt.% all examined similar electrochemical performances under atmospheric pressure, with an optimum at 2 wt.% PTFE. However, in a subsequent test series an electrolyte overpressure was applied. As a result, an electrolyte breakthrough accompanying with a performance loss at high current densities was observed for the 2 wt.% PTFE electrode. In contrast, the 6 wt.% PTFE electrode showed a better performance under the electrolyte overpressure for lower current densities. The amount of electrolyte inside the GDE was probably increased and thus the available three-phase surface area extended. However, a quantitative analysis of the electrolyte distribution and the reactive sides inside the GDE could not be achieved. The basic conclusion is that not the catalyst, but how it is utilized by the electrolyte is crucial.

Modelling the ORR in silver-based GDE

It is obvious, that a detailed description of the processes inside the GDE is only possible by modelling. The model used here is primarily based on the previous work of Pinnow et al. [36]. However, as mentioned in the introduction, the electrolyte transport inside the GDE was underestimated and local effects such as the solubility of oxygen were not captured adequately. A full summary on this topic is given in Table 4.1. Further, experimental data concerning the electrolyte distribution was not available at that time and the parameters were just adjusted to fit the polarization curves. Nevertheless, their values were quite good for a first estimation and close to the newly determined ones. With the upcoming operando experiments presented in chapter 4, a simple concept was developed to describe the distribution of the electrolyte fingers adequately (see Figure 4.2/6.1a and 4.3a). Combined with the spatial resolution of electrolyte properties, more detailed insights on the processes in the GDE are reached. However, the electrolyte distribution is still the key factor for the understanding. In the parameter study (see chapter 4.4.5), the influence of the distribution was already addressed, but only the impact on the polarization curve was considered. In this section, a more detailed analysis of the processes inside the GDE with only minor changes in the electrolyte distribution is discussed. In Figure 6.1a the parameter defining the electrolyte distribution are given again. In the industrial base case (31 wt.% NaOH, 100 vol.% O_2, see chapter 4) the flooded agglomerates have a radius of 5 µm, while the distance in between is 10 µm. Only one parameter is varied at a time and the interaction is not considered. Further, an electrolyte intrusion over the full thickness of the electrode is considered in all cases. Figure 6.1 summarizes the results from chapter 4.4.5.

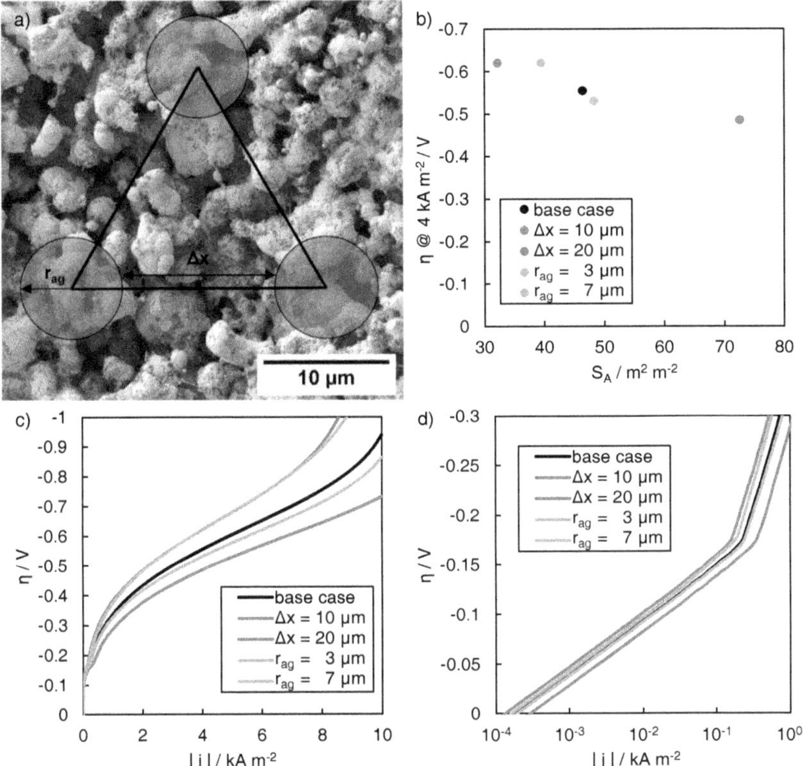

Figure 6.1: Influence of the electrolyte distribution on the polarization curve for the industrial base case (31 wt.% NaOH, 100 vol.% O_2). a) Front view of the flooded agglomerates and corresponding geometric parameters [38]. b) Resulting overpotential at industrial current density as a function of the surface enhancement factor. Polarization curves in linear (c) and semi-logarithmic (d) representation.

Figure 6.1b shows the resulting overpotential at 4 kA m^{-2} over the surface enhancement factor. Even with minor changes in the electrolyte distribution there is no linear correlation. An increase of the distance between single flooded agglomerates as well as a decrease in the agglomerate radius lead to higher overpotentials. In both cases ($\Delta x = 20\ \mu m$, $r_{ag} = 3\ \mu m$) the resulting polarization curves are almost identical. Increasing the radius to 7 µm slightly improves the performance. However, incipient mass transport limitations at 10 kA m^{-2} are still visible. Decreasing the distance to 10 µm boosts the electrodes performance significantly (see Figure 6.1c). In this case, the decrease of the overpotential is clearly visible even in the kinetic range.

Previously, the processes inside the GDE were only examined in detail for the industrial base case. The following analysis refers to the industrial current density of 4 kA m^{-2}. In Figure 6.2 the electrolyte accumulation and reaction rate inside the GDE for different electrolyte distributions is shown. As expected, the electrolyte distribution does not influence the concentration in the stagnant film, as the amount of hydroxide ions is only depends on the current density. On the interface between the film and the GDE the concentration reaches a value of 33.2 wt.% NaOH, 2.2 percentage points higher than in the electrolyte bulk. Inside the GDE the concentration profiles differ in dependence of the electrolyte distribution. Above a penetration depth of approx. 60 µm the curves split. However, the curves for the decreased distance ($\Delta x = 10\ \mu m$) and increased radius of flooded agglomerates ($r_{ag} = 7\ \mu m$) are almost identical. The concentration of the electrolyte at the end of the reaction layer is slightly higher than compared to the base case scenario. In contrast, a larger distance between the agglomerates leads to a reduction in the maximum concentration. However, the lowest values are reached with the smallest agglomerate radius. This is also reflected in the reaction rate. High concentration values at the end of the reaction zone go along with lower reaction rates at the corresponding position, due to the lower solubility of oxygen. To compensate this, the smallest flooded agglomerates will examine the highest reaction rates directly at the interface to the electrolyte. Near the electrolyte, the reaction rate is up to eleven times higher than near the gas compartment. With higher amounts of electrolyte inside the GDE ($\Delta x = 10\ \mu m$, $r_{ag} = 7\ \mu m$), the curve is flattened, having constant higher reaction rates near the gas compartment that only increase to approx. four times near the electrolyte interface. Here again, both curves are almost identical. For an intrusion depth of approx. 60 µm all curves overlap, which corresponds to the position where the electrolyte concentration starts to differ.

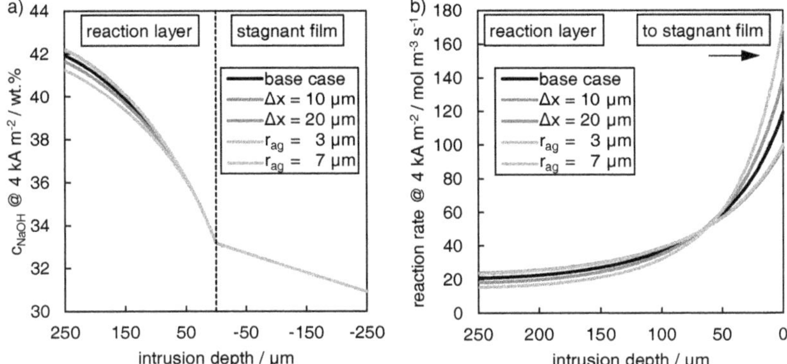

Figure 6.2: Electrolyte concentration (a) and reaction rate (b) as a function of the intrusion depth. Negative values of the intrusion depth represent the stagnant electrolyte film.

In Figure 6.3 a detailed look on the main fluxes inside the reaction layer is provided. The fluxes of oxygen are left out, as they correspond to the stoichiometric conversion to hydroxide ions (see Figure 4.7). Therefore, all effects concerning the oxygen flux can also be described by the hydroxide ion flux (Figure 6.3a). Higher amounts of electrolyte in the electrode (Δx = 10 µm, r_{ag} = 7 µm) leads to almost constant rise of the OH^- flux towards the electrolyte. In the other cases the reaction mainly takes place closer to the electrolyte interface. Therefore, the OH^- flux is lower for high intrusion depths and increase strongly towards the electrolyte compartment. This profiles are expected, as the OH^- (and O_2) flux is directly connected to the reaction rate (see eq. 4.8 and A8 in chapter 4). More interesting is the flux of water. Almost over the entire thickness of the electrode more electrolyte in the GDE (Δx = 10 µm, r_{ag} = 7 µm) leads to a higher flux of total water towards the gas compartment (see Figure 6.3b). Only at the interfaces towards the electrolyte and the gas compartment the flux is lower than compared to the other cases. This also results in a lower amount of condensate, if more electrolyte is inside the GDE. However, the difference is only 83 mg d^{-1} cm^{-2}, which is much smaller than fluctuations in measured values [39]. It is noticeable, that the profiles for $\Delta x = 10\ \mu m$ and $r_{ag} = 7\ \mu m$ are still almost identical. So far, no indication has been found as to why the polarization curves differ so much. On the other hand, the polarization curves for $\Delta x = 20\ \mu m$ and $r_{ag} = 3\ \mu m$ are almost equivalent, but all profiles in the detailed analysis are different. An examination of the profiles for the gaseous and liquid water transport (Figure 6.3c/d) will provide further information.

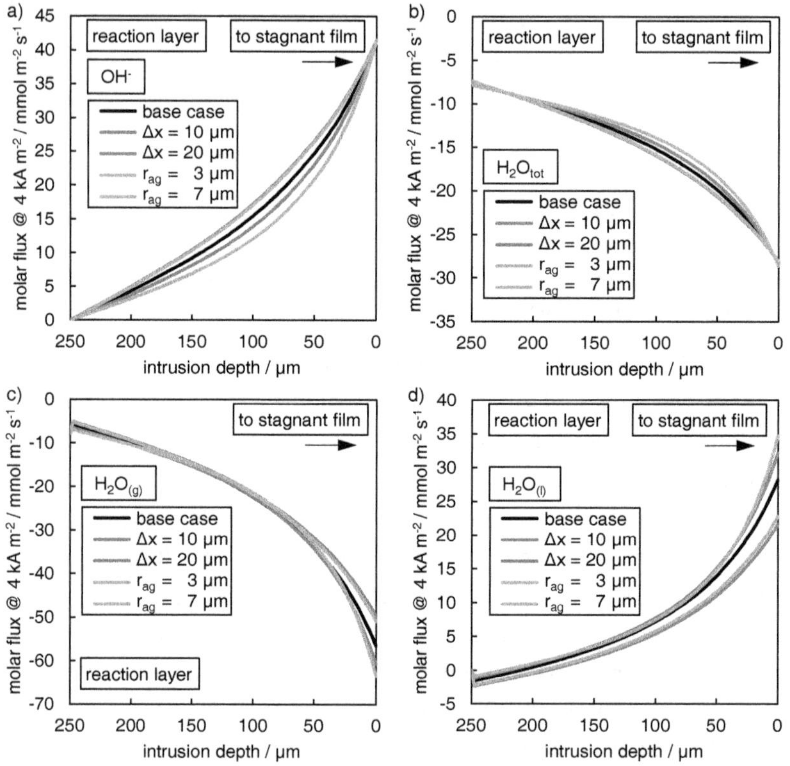

Figure 6.3: Molar fluxes of OH^- Ions (a), total water (b), as well as divided into gaseous (c) and liquid water (d) in the electrode as function of the intrusion depth. Positive fluxes direct towards the electrolyte, negative fluxes towards the gas compartment.

In contrast to the total water flux, more electrolyte inside the electrode will decrease the flux of gaseous water towards the gas compartment. Less electrolyte in the GDE increases the open pore space, and thus the flux of the gaseous species. Interestingly, the same is observed for the liquid water flux. Here, too, the flux is higher if less electrolyte is in the GDE. Increasing the amount of electrolyte in the GDE decreases the flux of liquid water towards the electrolyte compartment. In this case, a small deviation between the profiles of Δx = 10 µm and r_{ag} = 7 µm is observed. The flux for the Δx = 10 µm case is slightly smaller, so less water is needed to support the removal of the hydroxide ions from the reaction layer. This leads to a smaller transport resistance and could be a sign of why the performance in the polarization curve is much better. In contrast, the profiles of Δx = 20 µm and r_{ag} = 3 µm are almost identical, like the polarization curves. Another interesting observation is the direction of the flux. The liquid water is mainly transported towards the electrolyte in the flooded agglomerate. However, near the interface to the gas compartment the flux is reversed. The position of the reversal point depends on the electrolyte distribution. For the Δx = 10 µm case the reversal point is located at an intrusion depth of approx. 190 µm, while it is at 225 µm for the r_{ag} = 3 µm case. In the last 60 µm of the intrusion the removal of the ions is strongly inhibited, thus an accumulation takes place (see Figure 3.2a). This again highlights the importance of the effective removal of the hydroxide ions from the reaction layer supported by the flux of liquid water. Overall, the electrolyte distribution inside the GDE is a complex topic. Even small changes in the geometric parameters lead to dissimilar polarization curves and reaction, as well as transport profiles. If the electrolyte distribution is only adjusted on the basis of the polarization curve, this can lead to a misinterpretation on the further processes inside the GDE. Additional operando experiments are therefore fundamental. Still neglected in this analysis is the intrusion depth of the electrolyte. This topic has been discussed in detail in chapter 5, and is most likely in the range of half to full thickness of the electrode. To clarify this aspect, further operando tomography experiments are required.

Further possible improvements to the TFFA-model

Not yet included in the model are inhomogeneities in the electrode structure as well as dynamic effects that occur over time. In chapter 3 a complex pore network is revealed. In addition to that, Paulisch et al. [16] showed, that there are more pores of larger sizes inside the GDE. However, their function and share in the ORR are still unclear and are therefore not considered in the model. Further, only a static electrolyte distribution based on one operando experiment under room temperature is implemented. The implementation of a pore network model with electrolyte saturation [40] could allow a better estimation of the electrolyte distribution also for other electrode types in the future, all based on the capillary-flow porometry. However, before implementing such effects it has to be clarified how the geometric parameters defining the electrolyte distribution in the model are influenced by higher electrolyte saturation due to e.g. electrowetting [41]. Right now the distribution is defined by three parameters, the intrusion depth, radius and distance between the flooded agglomerates. The detailed analysis revealed, that a modification of only one of these parameters leads to complex changes of all processes inside the GDE. Therefore, it is important to validate these model changes decoupled from the electrochemical performance, similar to the approach followed in chapter 4. Besides further operando experiments, also with tomographies of the electrolyte distribution, capillary pressure saturation curves can be determined. The latter method is already established for fuel cells electrodes [42] and could also be extended to include electrowetting.

All the insights acquired are not only important for a more detailed description of the ORR in GDE, but also an adaptation to other reaction systems is possible. For example, the same electrodes can be used for the electrochemical reduction of CO_2 [43]. Here, too, related issues are addressed, such as local reaction environments [44]. Similar models describing transport and reaction inside the GDE are also being developed [45], but here the parameterization is even more challenging because the whole reaction system is more complex. Findings and methods from this thesis can also contribute to a better understanding of this system.

Appendix

A. GDE Model code

In this appendix, the complete gPROMS® code used for the simulations in Chapter 4 is revealed. The code was implemented and executed in the ModelBuilder version 6.0.4. Only abbreviations and indents have been introduced in some places to avoid unnecessary line breaks. A variation of this code is also available in the supporting information of the online version of chapter 5 [46].

Model Code

```
# Model to describe the Oxygen Reduction Reaction (ORR)
# in Gas Diffusion Electrode (GDE) for a given current
# density (i0) the overpotential (eta) is calculated

PARAMETER

GAS            AS ORDERED_SET DEFAULT['H2O','O2','N2']
ELECTROLYTE    AS ORDERED_SET DEFAULT['H2O_l','OH-','Na+']
COMPONENTE_LIST AS ORDERED_SET

#Constants

PI  AS REAL DEFAULT 3.14159265 # constant Pi
R   AS REAL DEFAULT 8.314  # Universal Gas c   J/(mol K)
F   AS REAL DEFAULT 96485        # Faraday c     As/mol
k_B AS REAL DEFAULT 1.3806504E-23 # Boltzmann c J/K

T0  AS REAL DEFAULT 273.15 # Kelvin temp. 0°C K
T_R AS REAL DEFAULT 298.15 # reference temp.  K

# Geometric electrode Properties

z_b            AS REAL # diff layer in GC          m
z_electrode AS REAL # electrode thickness       m
```

```
z_s          AS REAL # length gas diff layer    m
z_t          AS REAL # length reaction layer    m
z_d          AS REAL # el. diff layer in bulk   m

epsilon_s    AS REAL # porosity of electrode    1
epsilon_t    AS REAL #   gas in reaction layer  1
epsilon_n    AS REAL # el fraction in FA        1

tau_s        AS REAL # tortuosity of electrode  1
tau_t        AS REAL #   of gas channels        1
tau_n        AS REAL #   of FA                  1

r_s          AS REAL # radius of gas channels   m
r_t          AS REAL #   in reaction layer      m

r_ag         AS REAL # radius of FA             m
delta_tf     AS REAL # thickness of thin film   m
S_P_el       AS REAL # projection el surface    1
S_P_gas      AS REAL #   gas surface            1
S_tf         AS REAL # spec surface thin film   1/m
delta_x      AS REAL # distance between FA      m

# Kinetic Parameters

i_A          AS REAL # kinetic c 1. Tafel       Am³/m²mol
i_AA         AS REAL # kinetic c 2. Tafel       Am³/m²mol

T_s          AS REAL # 1. Tafel slope           V
T_ss         AS REAL # 2. Tafel slope           V
eta_Change   AS REAL # change 1 to 2 Tafel eq   V

#Material constants

M            AS ARRAY (COMPONENTE_LIST) OF REAL
                     # Molar mass               kg/mol
kappa_e      AS REAL # Conductivity Electrode   S/m

#Process conditions

T            AS REAL # Temperature              K
Theta        AS REAL # Temperature              °C

P_GC         AS ARRAY (GAS)      OF REAL
                     # Partial pressure in GC   Pa
r_PO2_GC     AS REAL # oxygen content in GC     1
rh_Feed      AS REAL # rel humidity in GC       1
```

```
P_tGC          AS REAL # Total pressure          Pa

# Diffusion Parameters

D12            AS ARRAY (GAS,GAS) OF REAL
               # Maxwell-Stefan diff coeff       m²/s
D12_s_eff      AS ARRAY (GAS,GAS) OF REAL
               # eff MS in diffusion layer       m²/s
D12_t_eff      AS ARRAY (GAS,GAS) OF REAL
               # eff MS in reaction layer        m²/s

D_K_s_eff      AS ARRAY (GAS)     OF REAL
               # eff Knudsen in diff layer       m²/s
D_K_t_eff      AS ARRAY (GAS)     OF REAL
               # eff. Knudsen in reaction layer m²/s

# Maxwell Stefan Diffusion Parameters

N              AS REAL
s              AS ARRAY (GAS)     OF REAL
e_k            AS ARRAY (GAS)     OF REAL
M_m            AS ARRAY (GAS,GAS) OF REAL
s_m            AS ARRAY (GAS,GAS) OF REAL
e_k_m          AS ARRAY (GAS,GAS) OF REAL
T_starr        AS ARRAY (GAS,GAS) OF REAL
Omega_D        AS ARRAY (GAS,GAS) OF REAL
O              AS ARRAY (1:8)     OF REAL
Diff_V         AS ARRAY (GAS)     OF REAL
AAA            AS ARRAY (1:2)     OF REAL

# Henry constant constants

H_k            AS REAL
H_kA           AS ARRAY (1:6)     OF REAL
H_phiA         AS ARRAY (1:3)     OF REAL

# vapor pressure constants

aa             AS ARRAY (1:11,1:3)OF REAL

# density constants

bb             AS ARRAY (1:6,1:3) OF REAL

# viscosity constants
eta_H2O        AS REAL # viscosity of water      Pa s
```

```
dd             AS ARRAY (1:5,1:3) OF REAL
ddd            AS ARRAY (1:5,1:2) OF REAL

# NaOH and H2O activity constants

NN             AS ARRAY (1:3,1:2) OF REAL
OO             AS ARRAY (1:8)     OF REAL
PP             AS ARRAY (1:3,1:2) OF REAL

# conductivity constants

CC             AS ARRAY (1:6)     OF REAL

# electrochemical standard potential parameters

E0             AS REAL # standard potential     V
E0_298         AS REAL DEFAULT 0.4011 # @25°C   V
dE0_dt         AS REAL DEFAULT -1.6816e-3 #     V/K
d2E0_dt2       AS REAL DEFAULT 7.23e-6    #     V/K²

 DISTRIBUTION_DOMAIN

zNodes_b AS [0:z_b] # domain for the gas compartment
zNodes_s AS [0:z_s] #  diffusion layer inside the GDE
zNodes_t AS [0:z_t] #  reaction layer inside the GDE
zNodes_d AS [0:z_d] #  liquid diff layer in the el bulk

 VARIABLE

# Diffusion layer in gas compartment

P_b      AS DISTRIBUTION (GAS,zNodes_b) OF Pressure
N_b      AS ARRAY        (GAS)          OF Molar_Flux

# Diffusion Layer in ODC Gas Diffusion Layer

P_s      AS DISTRIBUTION (GAS,zNodes_s) OF Pressure
N_s      AS ARRAY        (GAS)          OF Molar_Flux

# Diffusion Layer in ODC Reaction Layer

P_t     AS DISTRIBUTION (GAS, zNodes_t) OF Pressure
N_t     AS DISTRIBUTION (COMPONENTE_LIST, zNodes_t)
        OF Molar_Flux
         # molar fluxes in reaction layer     mol/m²s
N_t_H2Ot       AS DISTRIBUTION (zNodes_t) OF Molar_Flux
```

```
# Diffusion Layer in Electrolyte Bulk
N_d          AS ARRAY(ELECTROLYTE)         OF Molar_Flux
# gas properties
c            AS DISTRIBUTION (zNodes_t) OF Concentration
             # O2 concentration in el         mol/m³
ci           AS DISTRIBUTION (zNodes_t) OF Concentration
             # EQ O2 concentration in el      mol/m³
# reaction variables
rr           AS DISTRIBUTION (zNodes_t) OF Reaction_Rate
eta_eff      AS DISTRIBUTION (zNodes_t)
             OF Catalyst_Efficiency
Thiele       AS DISTRIBUTION (zNodes_t) OF Thiele_Module
k_c          AS DISTRIBUTION (zNodes_t)
             OF RRate_Constant
# electrolyte properties
w_NaOH       AS DISTRIBUTION (zNodes_t) OF Mass_Fraction
w_NaOH_d     AS DISTRIBUTION (zNodes_d) OF Mass_Fraction
m_NaOH       AS DISTRIBUTION (zNodes_t) OF Molality
m_NaOH_d     AS DISTRIBUTION (zNodes_d) OF Molality
c_NaOH       AS DISTRIBUTION (zNodes_t) OF Concentration
c_NaOH_d     AS DISTRIBUTION (zNodes_d) OF Concentration
x_NaOH       AS DISTRIBUTION (zNodes_t) OF Mole_fraction
x_NaOH_d     AS DISTRIBUTION (zNodes_d) OF Mole_fraction
M_NaOH_mean  AS DISTRIBUTION (zNodes_t)
             OF Molecular_weight
    # Henry constants
H            AS DISTRIBUTION (zNodes_t) OF Henryconstant
H_phi        AS DISTRIBUTION (zNodes_t) OF Help_Variable
    # densities
rho_NaOH     AS DISTRIBUTION (zNodes_t) OF Density
rho_NaOH_d   AS DISTRIBUTION (zNodes_d) OF Density
b        AS DISTRIBUTION (1:3,zNodes_t) OF Help_Variable
```

```
b_d       AS DISTRIBUTION (1:3,zNodes_d) OF Help_Variable

    #viscosities
eta_NaOH AS DISTRIBUTION (zNodes_t)     OF Viscosity
d        AS DISTRIBUTION (1:3,zNodes_t) OF Help_Variable

eta_NaOH_d  AS DISTRIBUTION (zNodes_d) OF Viscosity
d_d       AS DISTRIBUTION (1:3,zNodes_d) OF Help_Variable
    # vapor pressure
p_m         AS DISTRIBUTION (zNodes_t) OF Pressure
a         AS DISTRIBUTION (1:3,zNodes_t) OF Help_Variable

    # NaOH and H2O activity
gamma_NaOH  AS DISTRIBUTION (zNodes_t) OF Activity

a_H2O       AS DISTRIBUTION (zNodes_t) OF Activity
a_H2O_d     AS DISTRIBUTION (zNodes_d) OF Activity

    # conductivity
kappa_i0    AS DISTRIBUTION (zNodes_t) OF Conductivity
kappa_i     AS DISTRIBUTION (zNodes_t) OF Conductivity

# electrochemical properties

i0          AS Current_Density
eta         AS Potential

i_tf        AS DISTRIBUTION (zNodes_t)
            OF Current_Density
i_e         AS DISTRIBUTION (zNodes_t)
            OF Current_Density
i_i         AS DISTRIBUTION (zNodes_t)
            OF Current_Density

E_e         AS DISTRIBUTION (zNodes_t) OF Potential
E_i         AS DISTRIBUTION (zNodes_t) OF Potential
Delta_E     AS DISTRIBUTION (zNodes_t) OF Potential

    # standard potential

DeltaE_std  AS DISTRIBUTION (zNodes_t) OF Potential
ln_QT       AS DISTRIBUTION (zNodes_t) OF Help_Variable

# Diffusion coefficients                        m²/s

D_el_Na     AS DISTRIBUTION (zNodes_t) OF Diffusion
D_el_OH     AS DISTRIBUTION (zNodes_t) OF Diffusion
```

```
D_el_Na_eff AS DISTRIBUTION (zNodes_t) OF Diffusion
D_el_OH_eff AS DISTRIBUTION (zNodes_t) OF Diffusion

D_el_Na_d   AS DISTRIBUTION (zNodes_d) OF Diffusion
D_el_OH_d   AS DISTRIBUTION (zNodes_d) OF Diffusion

D_L_O2      AS DISTRIBUTION (zNodes_t) OF Diffusion
D_L_O2_eff  AS DISTRIBUTION (zNodes_t) OF Diffusion

backward    AS Help_Variable # for ini of the system
S_AAA       AS Help_Variable # enhancement   m²/m²

 SELECTOR

kappa_i0_mode AS (kappa_i0_ini, kappa_i0_real)
                 DEFAULT kappa_i0_real
D_el_Na_mode  AS (D_el_Na_ini,  D_el_Na_real)
                 DEFAULT D_el_Na_real
D_el_OH_mode  AS (D_el_OH_ini,  D_el_OH_real)
                 DEFAULT D_el_OH_real
D_L_O2_mode   AS (D_L_O2_ini,   D_L_O2_real)
                 DEFAULT D_L_O2_real

 SET
# Material constants
COMPONENTE_LIST := GAS + ELECTROLYTE;

M('H2O') := 18E-3; M('O2')     := 32E-3;
M('N2')  := 28E-3; M('H2O_l')  := 18E-3;
M('OH-') := 17E-3; M('Na+')    := 23E-3;

# definition of distribution

zNodes_b     := [BFDM, 2 , 12];
zNodes_s     := [BFDM, 2 , 12];
zNodes_t     := [BFDM, 2 , 108];
zNodes_d     := [BFDM, 2 , 12];

z_b          := 10e-3;

z_electrode  := 251e-6;
z_t          := 250e-6;
z_d          := 250e-6;
z_s          := z_electrode - z_t;

epsilon_s    := 0.38;
S_P_gas      := (1-(((PI/2)*(r_ag^2))
```

```
                / ((SQRT(3)/4)*((2*r_ag)+delta_x)^2)));
epsilon_t      := S_P_gas * epsilon_s;
epsilon_n      := (epsilon_s - epsilon_t)
                / (1 - epsilon_t);

tau_s          := 3.2;
tau_t          := 3.2;
tau_n          := 3.2;

r_s            := 0.35e-6;
r_t            := 0.35e-6;

r_ag           := 5e-6;
delta_x        := 15e-6;
delta_tf       := 20e-9;

S_P_el         := 2 * pi * r_ag^2
                / ( SQRT(3) * ( 2 * r_ag + delta_x)^2);
S_tf           := 2 * S_P_el * tau_n / r_ag;

# kinetic Parameters

i_A            := 0.1162;
i_AA           := i_A * (10^(eta_Change/T_s))
                      / (10^(eta_Change/T_ss));

T_s            := 0.054;
T_ss           := 0.210;

eta_Change     := 0.175;

# Process conditions

Theta          := 80;
T              := T0 + Theta;

P_tGC          := 1E5;
r_PO2_GC       := 1.0;
rh_Feed        := 0.05;

P_GC('H2O')    := 0.031699e5 * rh_Feed;  #100% rh @ 25°C
P_GC('O2')     := (P_tGC - P_GC('H2O'))
                * r_PO2_GC * 0.99; # 0.99 for impurities
P_GC('N2')     := P_tGC - P_GC('O2') - P_GC('H2O');

# Maxwell Stefan Diffusion Parameters
```

```

AAA(1) := 28.89e-10;  AAA(2) := 8.63e-10;

e_k('H2O') := 809.1;  s('H2O') := 2.641E-10;
e_k('O2')  := 106.7;  s('O2')  := 3.467E-10;
e_k('N2')  := 71.4;   s('N2')  := 3.798E-10;

Diff_V('H2O') := 13.1; Diff_V('O2'):= 16.3;
Diff_V('N2')  := 18.5;




O(1:8) := [1.06036,    0.15610,    0.19300,
            0.47635,    1.03587,    1.52996,
            1.76474,    3.89411];


N       := P_tGC / k_B / T;

  FOR i in GAS DO
      FOR j in GAS DO

      M_m(i,j)      := 2*((M(i)^(-1))+(M(j)^(-1)))^(-1);
      e_k_m(i,j)    := (e_k(i) * e_k(j))^(1/2);
      s_m(i,j)      := (s(i)+s(j))/2;
      T_starr(i,j)  := T / e_k_m(i,j);
      Omega_D(i,j)  := (O(1)/(T_starr(i,j)^O(2)))
                     + (O(3)/EXP(O(4)*T_starr(i,j)))
                     + (O(5)/EXP(O(6)*T_starr(i,j)))
                     + (O(7)/EXP(O(8)*T_starr(i,j)));
  D12(i,j) := (1e-3 * T^(1.75) * ((M_m(i,j))^(1/2))
            / ((P_tGC/1.01325E5) * ((Diff_V(i)^(1/3))
            + (Diff_V(j)^(1/3)))^2)) / 10000;
      END
  END

D12_s_eff := D12 * epsilon_s / tau_s;
D12_t_eff := D12 * epsilon_t / tau_t;

D_K_s_eff := epsilon_s / tau_s * 2/3
           * r_s * (8 * R * T / (PI * M(GAS)))^0.5;
D_K_t_eff := epsilon_t / tau_t * 2/3
           * r_t * (8 * R * T / (PI * M(GAS)))^0.5;

# for Henry constant calculations
```

```
H_kA(1) :=    298.0;      H_phiA (1) := 0.102078;
H_kA(2) :=      0.046;    H_phiA (2) := 1.00044;
H_kA(3) :=    203.35;     H_phiA (3) := 4.308933;
H_kA(4) :=    299.378;
H_kA(5) :=      0.092;
H_kA(6) := -20591.0;

H_k := EXP(((H_kA(2)*(T^2)) + (H_kA(3) * T *
LOG(T/H_kA(1))) - ((H_kA(4) + H_kA(5) * T) * (T -
H_kA(1))) + H_kA(6)) / (R * T));

# for vapor pressure of H2O in NaOH calculations

aa( 1,1) :=   -113.93947;  aa( 1,2) :=    16.240074;
aa( 1,3) :=   -226.80157;
aa( 2,1) :=    209.82305;  aa( 2,2) :=    -11.864008;
aa( 2,3) :=    293.17155;
aa( 3,1) :=    494.77153;  aa( 3,2) :=   -223.47305;
aa( 3,3) :=   5081.8791;
aa( 4,1) :=   6860.8330;   aa( 4,2) :=  -1650.3997;
aa( 4,3) :=  36752.126;
aa( 5,1) :=   2676.643;    aa( 5,2) :=  -5997.3118;
aa( 5,3) := 131262.00;
aa( 6,1) := -21740.328;    aa( 6,2) := -12318.744;
aa( 6,3) := 259399.54;
aa( 7,1) := -34750.872;    aa( 7,2) := -15303.153;
aa( 7,3) := 301696.22;
aa( 8,1) := -20122.157;    aa( 8,2) := -11707.480;
aa( 8,3) := 208617.90;
aa( 9,1) :=  -4105.9890;   aa( 9,2) :=  -5364.9554;
aa( 9,3) :=  81774.024;
aa(10,1) :=      0;        aa(10,2) :=  -1338.5412;
aa(10,3) :=  15648.526;
aa(11,1) :=      0;        aa(11,2) :=   -137.96889;
aa(11,3) :=    906.29769;

# for density calculation

bb(1,1) :=    5007.2279636;  bb(1,2) :=   -
64.786269079;       bb(1,3) := 0.24436776978;
bb(2,1) :=  -25131.164248;   bb(2,2) :=   525.34360564;
bb(2,3) := -1.9737722344;
bb(3,1) :=   74107.692582;   bb(3,2) := -1608.4471903;
bb(3,3) :=  6.04601497138;
bb(4,1) := -104657.48692582; bb(4,2) :=  2350.9753235;
bb(4,3) := -8.9090614947;
bb(5,1) :=   69821.773186;   bb(5,2) := -1660.9035108;
bb(5,3) :=  6.37146769397;
```

```
bb(6,1) := -18145.911810;   bb(6,2) :=   457.64374355;
bb(6,3) := -1.7816083111;
# for viscosity calculations
dd(1,1) :=   -6.1420727; dd(1,2) :=   2.3171396;
dd(1,3) := -0.1152143;
dd(2,1) :=  124.64849;   dd(2,2) := -23.153644;
dd(2,3) :=  1.0543467;
dd(3,1) := -247.08170;   dd(3,2) :=  49.267937;
dd(3,3) := -2.3693277;
dd(4,1) :=  147.73585;   dd(4,2) := -36.970260;
dd(4,3) :=  2.0099091;
dd(5,1) :=    0;         dd(5,2) :=   6.5882887;
dd(5,3) := -0.5257284;
ddd(1,1) :=  0.5868156;  ddd(1,2) := 0;
ddd(2,1) := -0.0398182;  ddd(2,2) := 1;
ddd(3,1) :=  0.00247793; ddd(3,2) := 1.5;
ddd(4,1) := -4.9427e-6;  ddd(4,2) := 2.5;
ddd(5,1) :=  1.48701e-7; ddd(5,2) := 3;
eta_H2O := 1e-3 * EXP(SIGMA(ddd(1,1)* Theta^ddd(1,2),
ddd(2,1)* Theta^ddd(2,2),   ddd(3,1)* Theta^ddd(3,2),
ddd(4,1)* Theta^ddd(4,2),   ddd(5,1)* Theta^ddd(5,2)));
# for NaOH and H2O activity calculations
NN(1,1) := -0.01332;  OO(1) := -139.958399;
PP(1,1) := 0.141102442;
NN(2,1) :=  0.002542; OO(2) :=  176.294068;
PP(2,1) := 0.001273918392;
NN(3,1) := -3.06e-5;  OO(3) :=   -2.98871232;
PP(3,1) := 0.14876612;
                      OO(4) :=    0.0208479756;
                      OO(5) :=   -7.65562053e-5;
NN(1,2) :=  1.5827;   OO(6) :=    1.5619109e-7;
PP(1,2) := -86.97615589;
NN(2,2) := -1.5669;   OO(7) :=   -1.68003492e-10;
PP(2,2) :=  -0.886580916;
NN(3,2) :=  0.021296; OO(8) :=    7.44709291e-14;
PP(3,2) := 101.4477;
# for NaOH conductivity calculations
CC(1) :=    0.8041;
CC(2) := -  8.5203;
CC(3) :=  123.3;
```

```
CC(4) := -548.15;
CC(5) :=  997.56;
CC(6) := -651.05;
# Electrode conductivity
kappa_e := 1e5;
# standard potential calculations
E0 := E0_298 + (T - T_R) * dE0_dt
    + 0.5 * (T - T_R)^2 * d2E0_dt2;
 BOUNDARY
# Boundary condition equations
N_s('N2')    = 0;      # no N2 generated or consumed
#transfer from GC to diff layer in GC
P_GC(GAS)    = P_b(GAS,0);
#transfer form diff layer in GC to diff layer in ODC
N_b          = N_s;    # transfer fluxes
P_b(,z_b)    = P_s(,0);#     partial pressures
#transfer from diff layer in ODC to reaction layer
N_s(GAS)         = N_t(GAS,0);  # transfer fluxes
P_s(GAS,z_s)     = P_t(GAS,0);  #    partial pressures
# Boundary conditions for the reaction layer
FOR z:= 0 TO z_t DO
N_t_H2Ot(z) = N_t('H2O',z) + N_t('H2O_l',z);
END
N_t('H2O_l',0)       = 0; # no l water in gas diff layer
N_t('O2',z_t)        = 0; # no gas in el phase
N_t('N2',0|+:z_t)    = 0; # no reaction of N2
N_t('Na+',)          = 0; # no reaction of Na+
# pressure of H2O is equal to vapor pressure of el
FOR z:= 0 TO z_t DO
P_t('H2O',z) = p_m(z);
END
# conditions interface reaction and liquid diff layer
N_d('Na+')       = N_t('Na+',z_t);
N_d('OH-')       = N_t('OH-',z_t);
```

```
N_d('H2O_l')      = N_t_H2Ot(z_t);

w_NaOH(z_t)       = w_NaOH_d(0);

# Boundary conditions for electrolyte potential
E_i(0) = 0;


 EQUATION

# dynamic 'experiment': increase of current density
# --> steady state for all other variables!
IF backward = -1 THEN
$i0 = -10; # backward loop to reach kinetic region
ELSE
$i0 = 100; # going from 0 to 10 kA/m²
END

# gas transport in diffusion layer in gas chamber
N_b('O2') = i0 / ( 4 * F);

For i IN GAS -'N2' DO
    FOR z:= 0|+ TO z_b DO
    (-1 / (R * T)) * PARTIAL(P_b(i,z),zNodes_b)
    = INTEGRAL( j over GAS;
      (P_b(j,z) * N_b(i) - P_b(i,z) * N_b(j))
    / (SIGMA(P_b(GAS,z)) * D12(i,j)));
    END
END
FOR z := 0|+ TO z_b DO
P_tGC = SIGMA(P_b(GAS,z));
END

# gas transport in ODC diffusion layer

For i IN GAS -'N2' DO
    FOR z:= 0|+ TO z_s DO
    (-1 / (R * T)) * PARTIAL(P_s(i,z),zNodes_s)
    = N_s(i) / D_K_s_eff(i) + INTEGRAL( j over GAS;
      (P_s(j,z) * N_s(i) - P_s(i,z) * N_s(j))
    / (SIGMA(P_s(GAS,z)) * D12_s_eff(i,j)));
    END
END
FOR z := 0|+ TO z_s DO
P_tGC = SIGMA(P_s(GAS,z));
END
```

```
#------------------------------------------------------------
# reaction layer -------------------------------------------
#------------------------------------------------------------

    # gas transport in ODC reaction layer


For i IN GAS - 'N2' DO
    FOR z:= 0|+ TO z_t DO
    (-1 / (R * T)) * PARTIAL(P_t(i,z),zNodes_t)
    = N_t(i,z) / D_K_t_eff(i) + INTEGRAL( j over GAS;
      (P_t(j,z) * N_t(i,z) - P_t(i,z) * N_t(j,z))
    / (SIGMA(P_t(GAS,z)) * D12_t_eff(i,j)));
    END
END
FOR z := 0|+ TO z_t DO
P_tGC = SIGMA(P_t(GAS,z));
END

FOR z:= 0|+ TO z_t|- DO # stoichiometric reac of water
N_t_H2Ot(z)     = N_t_H2Ot(z_t) + 2 * N_t('O2',z);
END
                         # global mass balance
N_t_H2Ot(0)     = N_t_H2Ot(z_t) + 2 * N_b('O2');

FOR z:= 0 TO z_t|- DO   # stoichiometric reac to OH-
N_t('OH-',z) = N_t('OH-',z_t) - 4 * N_t('O2',z);
END
N_t('OH-',z_t) = 4 * N_b('O2');

# MS diff of liquid water inside the reaction layer
FOR z:= 0|+ TO z_t DO
- PARTIAL(LOG(a_H2O(z)), zNodes_t)
  = (((((x_NaOH(z)) * N_t('H2O_l',z))
  - ((1 - x_NaOH(z))*N_t('OH-',z))) / D_el_OH_eff(z))
  + (((x_NaOH(z)) * N_t('H2O_l',z)) / D_el_Na_eff(z)))
  / (((1 - w_NaOH(z)) * rho_NaOH(z)) / M('H2O'));
END

# # transport in electrolyte bulk
FOR z:= 0|+ TO z_d DO

- PARTIAL(LOG(a_H2O_d(z)), zNodes_d)
  = (( x_NaOH_d(z) * N_d('H2O_l')
  - (1 - x_NaOH_d(z)) * N_d('OH-') ) / D_el_OH_d(z)
```

```
  + (x_NaOH_d(z) * N_d('H2O_l'))     / D_el_Na_d(z))
  / ((1 - w_NaOH_d(z)) * rho_NaOH_d(z) / M('H2O'));
END

# electrochemical reaction
    # Tafel kinetics for the reaction
FOR z := 0 TO z_t DO
IF Delta_E(z) < eta_Change THEN

i_tf(z) = i_A *   c(z) * 10^(Delta_E(z)/T_s);
ELSE
i_tf(z) =  i_AA * c(z) * 10^(Delta_E(z)/T_ss);
END
END

    # description of the reaction rate
FOR z := 0|+ TO z_t DO # RR by conversion of oxygen
rr(z) = - PARTIAL(N_t('O2',z), zNodes_t);
END

rr = (i_tf / ( 4 * F)) * S_tf; # RR by tf current
rr = (D_L_O2 / delta_tf) * ( ci - c) * S_tf; # O2 diff
rr = k_c * c * eta_eff * (1- epsilon_t); # reac in FA

    #Thiele module for reaction
    4 * Thiele^2 = r_ag^2 * (k_c / D_L_O2);
    FOR z := 0 TO z_t DO
    IF Thiele(z) >= 15 THEN # 15 for stability
    eta_eff(z) * Thiele(z) = 1 ;
    ELSE
    eta_eff(z) * Thiele(z)   = TANH(Thiele(z));
    END
    END

# potential difference of electrode and electrolyte
# as driving force for electrochemical reaction
Delta_E = E_e - E_i - DeltaE_std;

# calculating the resulting overpotential
eta = E_e(0) - E_i(z_t) - DeltaE_std(z_t);

# Potentials according to ohms law
FOR z := 0|+ TO z_t DO # pot dist inside electrode
PARTIAL(E_e(z),zNodes_t)     = - i_e(z) / kappa_e;
END

```

```
FOR z := 0|+ TO z_t DO # pot dist inside electrolyte
PARTIAL(E_i(z),zNodes_t)     = - i_i(z) / kappa_i(z);
END

FOR z := 0|+ TO z_t DO
i_e(z) = 4 * F * N_t('O2',z);
END
i0  = i_e + i_i;
i0  = i_e(0);

## Diffusion coefficients
# MS diffusion coefficients in liquid phase
CASE D_el_Na_mode OF # initialize system
    WHEN D_el_Na_ini:
    D_el_Na           =   8.2165E-10 ;
    D_el_Na_d         =   8.2165E-10 ;
    WHEN D_el_Na_real:
    D_el_Na           =   k_B * T /(eta_NaOH * AAA(1));
    D_el_Na_d         =   k_B * T /(eta_NaOH_d * AAA(1));
END

CASE D_el_OH_mode OF # initialize system
    WHEN D_el_OH_ini:
    D_el_OH           =   8.2165E-10 ;
    D_el_OH_d         =   8.2165E-10 ;
    WHEN D_el_OH_real:
    D_el_OH           =   k_B * T /(eta_NaOH * AAA(2));
    D_el_OH_d         =   k_B * T /(eta_NaOH_d * AAA(2));
END

# diffusion coefficient of oxygen in electrolyte
# calculated by modified Wilke and Chang equation
CASE D_L_O2_mode OF # initialize system
    WHEN D_L_O2_ini:
    D_L_O2 =   2.21294E-09;
    WHEN D_L_O2_real:
D_L_O2 = k_B * T / (6 * PI * 72e-12 * (eta_NaOH^0.36))
       * (1000/1000^0.36);
END

D_el_Na_eff    = D_el_Na * epsilon_n / tau_n;
D_el_OH_eff    = D_el_OH * epsilon_n / tau_n;

D_L_O2_eff     = D_L_O2 * epsilon_n / tau_n;
```

```
# simple conversions

w_NaOH       = m_NaOH   *   (M('Na+') + M('OH-'))
             / (1 + m_NaOH*    (M('Na+') + M('OH-')));
w_NaOH_d     = m_NaOH_d *   (M('Na+') + M('OH-'))
             / (1 + m_NaOH_d* (M('Na+') + M('OH-')));

c_NaOH       = m_NaOH   * rho_NaOH
             / (1 + (m_naOH *    (M('Na+') + M('OH-'))));
c_NaOH_d     = m_NaOH_d * rho_NaOH_d
             / (1 + (m_naOH_d * (M('Na+') + M('OH-'))));

x_NaOH       = (w_NaOH   / (M('Na+') + M('OH-')))
             * ( 1 /((w_NaOH   / (M('Na+') + M('OH-')))
             + ((1 - w_NaOH)   / M('H2O'))));
x_NaOH_d     = (w_NaOH_d / (M('Na+') + M('OH-')))
             * ( 1 /((w_NaOH_d / (M('Na+') + M('OH-')))
             + ((1 - w_NaOH_d) / M('H2O'))));

M_NaOH_mean = (1/((w_NaOH/(M('Na+') + M('OH-')))
             + ((1-w_NaOH)/M('H2O'))))*1000 ;

# Henry constant calculations

H_phi   = (1/ (1+ H_phiA(1) * m_NaOH^H_phiA(2)))
         ^ H_phiA(3);
H       = H_k * rho_NaOH * H_phi / 1.01325E5;

ci / H = P_t('O2',);

# vapor pressure calculations

FOR z := 0 TO z_t DO
    FOR i := 1 TO 3 DO
    a(i,z) = INTEGRAL(j := 1:11;
                 aa(j,i) * (LOG(1-w_NaOH(z)))^(j-1));
    END
END
p_m = 1000 * EXP((a(1,) + a(2,)*Theta)/(Theta-a(3,)));

# density calculation

FOR z := 0 to z_t DO
    FOR i := 1 to 3 DO
    b(i,z)   = SIGMA(bb(1,i) * (1-w_NaOH(z))^(0/2),
                     bb(2,i) * (1-w_NaOH(z))^(1/2),
```

```
                    bb(3,i) * (1-w_NaOH(z))^(2/2),
                    bb(4,i) * (1-w_NaOH(z))^(3/2),
                    bb(5,i) * (1-w_NaOH(z))^(4/2),
                    bb(6,i) * (1-w_NaOH(z))^(5/2));
    END
rho_NaOH(z) = SIGMA(b(1,z)* Theta^0, b(2,z)* Theta^1,
                    b(3,z)* Theta^2 );
END

FOR z := 0 to z_d DO
    FOR i := 1 to 3 DO
    b_d(i,z)  = SIGMA(bb(1,i) * (1-w_NaOH_d(z))^(0/2),
                      bb(2,i) * (1-w_NaOH_d(z))^(1/2),
                      bb(3,i) * (1-w_NaOH_d(z))^(2/2),
                      bb(4,i) * (1-w_NaOH_d(z))^(3/2),
                      bb(5,i) * (1-w_NaOH_d(z))^(4/2),
                      bb(6,i) * (1-w_NaOH_d(z))^(5/2));
    END
rho_NaOH_d(z) = SIGMA(b_d(1,z)* Theta^0,
                      b_d(2,z)* Theta^1,
                      b_d(3,z)* Theta^2);
END

# viscosity calculation

FOR z := 0 to z_t DO

d(1,z) = INTEGRAL(i := 1:5; dd(i,1) * w_NaOH(z)^i);
d(2,z) = INTEGRAL(i := 1:5; dd(i,2) * w_NaOH(z)^i);
d(3,z) = INTEGRAL(i := 1:5; dd(i,3) * w_NaOH(z)^i);

eta_NaOH(z) = eta_H2O * EXP( SIGMA(d(1,z)*Theta^(0/2),
              d(2,z)*Theta^(1/2), d(3,z)*Theta^(2/2)));
END

FOR z := 0 to z_d DO

d_d(1,z) = INTEGRAL(i := 1:5; dd(i,1) * w_NaOH_d(z)^i);
d_d(2,z) = INTEGRAL(i := 1:5; dd(i,2) * w_NaOH_d(z)^i);
d_d(3,z) = INTEGRAL(i := 1:5; dd(i,3) * w_NaOH_d(z)^i);

eta_NaOH_d(z) = eta_H2O * EXP(
 SIGMA(d_d(1,z)*Theta^(0/2), d_d(2,z)*Theta^(1/2),
       d_d(3,z)*Theta^(2/2)));
END
```

```
# NaOH and H2O Activity calculation
gamma_NaOH = 10^(INTEGRAL( i := 1:8; OO(i)*T^(i-2))
            - ( PP(1,1) + PP(1,2)/T ) * m_NaOH
            + ( PP(2,1) + PP(2,2)/T ) * m_NaOH^2
            - ( PP(3,1) + PP(3,2)/T ) * log10(m_NaOH));
a_H2O    = 10^(INTEGRAL(i := 1:3; NN(i,1) * m_NaOH^i)
         + INTEGRAL(i := 1:3; NN(i,2) * m_NaOH^i)/T);
a_H2O_d = 10^(INTEGRAL(i := 1:3; NN(i,1) * m_NaOH_d^i)
         + INTEGRAL(i := 1:3; NN(i,2) * m_NaOH_d^i)/T);
# Conductivity calculations, only valid for 80°C
CASE kappa_i0_mode OF # initialize system
    WHEN kappa_i0_ini:
    kappa_i0 = 103.19;
    WHEN kappa_i0_real:
    kappa_i0 = (INTEGRAL(i := 1:6;
                CC(i)*w_NaOH^(i-1)))*100;
END
kappa_i      = epsilon_n / tau_n * kappa_i0;
# standard potential calculations
ln_QT        = LOG(((gamma_NaOH * m_NaOH)^4)
             / (((ci/1000) * (a_H2O^2))));
DeltaE_std   = E0 - ( (R * T) / (4 * F)) * ln_QT;
S_AAA        = S_tf * z_t;
 ASSIGN
w_NaOH_d(z_d) := 0.3091; #NaOH bulk concentration
backward := -1;
 PRESET
P_s('H2O',) := 0.2e5;
P_b('H2O',) := 0.2e5;
P_t('H2O',) := 0.2e5;
P_b('N2',)  := 0.1e5 : -10000:300000;
```

```
P_s('N2',)  := 0.1e5 : -10000:300000;
P_t('N2',)  := 0.1e5 : -10000:300000;

N_t('H2O_l',)       := -0.01;

w_NaOH      := 0.30;
x_NaOH      := 0.18;

a_H2O       := 0.53;
c_NaOH      := 10000;

D_el_Na     := 8e-10;
D_el_OH     := 3e-09;

a(1,)       := -350;
a(2,)       :=   16;
a(3,)       := -237;

c           := 3e-2;
ci          := 3.1e-2;
H           := 4.2e-7;

E_e         := 0.3;
E_i         := 0;

Delta_E     := 0.3;
i_tf        := 0.1 : 0 : 15000;

gamma_NaOH := 1.8;
ln_QT       := 19.65;


INITIALISATION_PROCEDURE Initial_1
START
kappa_i0_mode    := kappa_i0_ini;
D_el_Na_mode     := D_el_Na_ini;
D_el_OH_mode     := D_el_OH_ini;
D_L_O2_mode      := D_L_O2_ini;
END
NEXT
    MOVE_TO
        kappa_i0_mode    := kappa_i0_real;
    END
END

NEXT
```

```
    MOVE_TO
        D_el_Na_mode      := D_el_Na_real;
    END
END

NEXT
    MOVE_TO
        D_el_OH_mode      := D_el_OH_real;
    END
END
NEXT
    MOVE_TO
        D_L_O2_mode       := D_L_O2_real;
    END
END
```

Process Code

```
# Process to start the ODC model

 PARAMETER

Output AS ORDERED_SET DEFAULT ['j / kA/m²','eta / V']

 UNIT
ODC      AS ODC

 VARIABLE

Pola            AS ARRAY (Output)  OF Help_Variable
m_condensate AS                       Help_Variable

 EQUATION
#write an array for the polarization curve
Pola('j / kA/m²')   = ODC.i0/1000;
Pola('eta / V')     = -ODC.eta;

#expected condensate
m_condensate = -ODC.N_b('H2O') * ODC.M('H2O')
               * 1000 * (0.01^2) * 3600 * 24;

INITIAL
ODC.i0 = 10;

INITIALISATION_PROCEDURE
```

```
# Start Default Initialization Procedure
    USE
                : DEFAULT;
    END
    SAVE "ODC_InitialGuess";

 SCHEDULE
# start with kinetic region
SEQUENCE
CONTINUE FOR 0.99

# start with 0 to 10 kA/m²
REASSIGN
ODC.backward := 0;
END
CONTINUE FOR 100
END
```

References

[1] UNFCCC, ADOPTION OF THE PARIS AGREEMENT - Paris Agreement text English, https://unfccc.int/sites/default/files/english_paris_agreement.pdf accessed 29 March 2021.

[2] A. Boulamanti, J.A. Moya, Energy efficiency and GHG emissions: Prospective scenarios for the Chemical and Petrochemical Industry. https://doi.org/10.2760/20486.

[3] J. Pérez-Ramírez, C. Mondelli, T. Schmidt, O.F.-K. Schlüter, A. Wolf, L. Mleczko, T. Dreier, Sustainable chlorine recycling via catalysed HCl oxidation: from fundamentals to implementation, Energy Environ. Sci. 4 (2011) 4786. https://doi.org/10.1039/c1ee02190g.

[4] J. Kintrup, M. Millaruelo, V. Trieu, A. Bulan, E.S. Mojica, Gas Diffusion Electrodes for Efficient Manufacturing of Chlorine and Other Chemicals, Electrochem. Soc. Interface 26 (2017) 73–76. https://doi.org/10.1149/2.F07172if.

[5] IEA (2020), Electricity Information: Overview: IEA, Paris, https://www.iea.org/reports/electricity-information-overview, accessed 29 March 2021.

[6] Hydrogen from chlor-alkali production: as green as green can be, 2021, https://www.eurochlor.org/news/hydrogen-from-chlor-alkali-production-as-green-as-green-can-be/, accessed 15 March 2021.

[7] I. Moussallem, S. Pinnow, T. Turek, Möglichkeiten zur Energierückgewinnung aus Wasserstoff bei der Chlor-Alkali-Elektrolyse, Chem. Ing. Tech. 81 (2009) 489–493. https://doi.org/10.1002/cite.200800172.

[8] Chlor-Alkali Electrolyis, https://ucpcdn.thyssenkrupp.com/_legacy/UCPthyssenkruppBAISUhdeChlorineEngineers/assets.files/products/chlor_alkali_electrolysis/thyssenkrupp_chlor_alkali_brochure_web.pdf, accessed 22 October 2020.

[9] D. Öhl, D. Franzen, M. Paulisch, S. Dieckhöfer, S. Barwe, C. Andronescu, I. Manke, T. Turek, W. Schuhmann, Catalytic Reactivation of Industrial Oxygen Depolarized Cathodes by in situ Generation of Atomic Hydrogen, ChemSusChem 12 (2019) 2732–2739. https://doi.org/10.1002/cssc.201900628.

[10] J. Weigert, C. Hoffmann, E. Esche, P. Fischer, J.-U. Repke, Towards demand-side management of the chlor-alkali electrolysis: Dynamic modeling and model validation, Comput. Chem. Eng. 149 (2021) 107287. https://doi.org/10.1016/j.compchemeng.2021.107287.

[11] D. Tromans, Oxygen solubility modeling in inorganic solutions: concentration, temperature and pressure effects, Hydrometallurgy 50 (1998) 279–296. https://doi.org/10.1016/S0304-386X(98)00060-7.

[12] S.G. Bratsch, Standard Electrode Potentials and Temperature Coefficients in Water at 298.15 K, J. Phys. Chem. Ref. Data 18 (1989) 1–21. https://doi.org/10.1063/1.555839.

[13] J. Jörissen, T. Turek, R. Weber, Chlorherstellung mit Sauerstoffverzehrkathoden. Energieeinsparung bei der Elektrolyse, Chem. unserer Zeit 45 (2011) 172–183. https://doi.org/10.1002/ciuz.201100545.

[14] I. Moussallem, J. Jörissen, U. Kunz, S. Pinnow, T. Turek, Chlor-alkali electrolysis with oxygen depolarized cathodes: history, present status and future prospects, J. Appl. Electrochem. 38 (2008) 1177–1194. https://doi.org/10.1007/s10800-008-9556-9.

[15] N. Furuya, H. Aikawa, Comparative study of oxygen cathodes loaded with Ag and Pt catalysts in chlor-alkali membrane cells, Electrochim. Acta 45 (2000) 4251–4256. https://doi.org/10.1016/S0013-4686(00)00557-0.

[16] M.C. Paulisch, M. Gebhard, D. Franzen, A. Hilger, M. Osenberg, N. Kardjilov, B. Ellendorff, T. Turek, C. Roth, I. Manke, Operando Laboratory X-Ray Imaging of Silver-Based Gas Diffusion Electrodes during Oxygen Reduction Reaction in Highly Alkaline Media, Materials 12 (2019). https://doi.org/10.3390/ma12172686.

[17] M. Neumann, M. Osenberg, A. Hilger, D. Franzen, T. Turek, I. Manke, V. Schmidt, On a pluri-Gaussian model for three-phase microstructures, with applications to 3D image data of gas-diffusion electrodes, Comput. Mater. Sci. 156 (2019) 325–331. https://doi.org/10.1016/j.commatsci.2018.09.033.

[18] P. Kunz, M. Hopp-Hirschler, U. Nieken, Simulation of Electrolyte Imbibition in Gas Diffusion Electrodes, Chem. Ing. Tech. 91 (2019) 883–888. https://doi.org/10.1002/cite.201800202.

[19] P. Kunz, M. Paulisch, M. Osenberg, B. Bischof, I. Manke, U. Nieken, Prediction of Electrolyte Distribution in Technical Gas Diffusion Electrodes: From Imaging to SPH Simulations, Transp. Porous Med. 132 (2020) 381–403. https://doi.org/10.1007/s11242-020-01396-y.

[20] F. Kubannek, T. Turek, U. Krewer, Modeling Oxygen Gas Diffusion Electrodes for Various Technical Applications, Chem. Ing. Tech. 91 (2019) 720–733. https://doi.org/10.1002/cite.201800181.

[21] Y. Wang, Modeling discharge deposit formation and its effect on lithium-air battery performance, Electrochim. Acta 75 (2012) 239–246. https://doi.org/10.1016/j.electacta.2012.04.137.

[22] M. Röhe, F. Kubannek, U. Krewer, Processes and Their Limitations in Oxygen Depolarized Cathodes: A Dynamic Model-Based Analysis, ChemSusChem 12 (2019) 2373–2384. https://doi.org/10.1002/cssc.201900312.

[23] M. Röhe, A. Botz, D. Franzen, F. Kubannek, B. Ellendorff, D. Öhl, W. Schuhmann, T. Turek, U. Krewer, The Key Role of Water Activity for the Operating Behavior and Dynamics of Oxygen Depolarized Cathodes, ChemElectroChem 6 (2019) 5671–5681. https://doi.org/10.1002/celc.201901224.

[24] R.R. Sijabat, M.T. de Groot, S. Moshtarikhah, J. van der Schaaf, Maxwell–Stefan model of multicomponent ion transport inside a monolayer Nafion membrane for intensified chlor-alkali electrolysis, J. Appl. Electrochem. 49 (2019) 353–368. https://doi.org/10.1007/s10800-018-01283-x.

[25] N.S. Vasile, R. Doherty, A.H.A. Monteverde Videla, S. Specchia, 3D multi-physics modeling of a gas diffusion electrode for oxygen reduction reaction for electrochemical energy conversion in PEM fuel cells, Appl. Energy 175 (2016) 435–450. https://doi.org/10.1016/j.apenergy.2016.04.030.

[26] L.G. Austin, M. Ariet, R.D. Walker, G.B. Wood, R.H. Comyn, Simple-Pore and Thin-Film Models of Porous Gas Diffusion Electrodes, Ind. Eng. Chem. Fund. 4 (1965) 321–327. https://doi.org/10.1021/i160015a015.

[27] S. Srinivasan, H.D. Hurwitz, Theory of a thin film model of porous gas-diffusion electrodes, Electrochim. Acta 12 (1967) 495–512. https://doi.org/10.1016/0013-4686(67)80019-7.

[28] R.P. Iczkowski, Mechanism of the Hydrogen Gas Diffusion Electrode, J. Electrochem. Soc. 111 (1964) 1078. https://doi.org/10.1149/1.2426320.

[29] F.G. Will, Electrochemical Oxidation of Hydrogen on Partially Immersed Platinum Electrodes, J. Electrochem. Soc. 110 (1963) 145. https://doi.org/10.1149/1.2425692.

[30] F.G. Will, Electrochemical Oxidation of Hydrogen on Partially Immersed Platinum Electrodes, J. Electrochem. Soc. 110 (1963) 152. https://doi.org/10.1149/1.2425693.

[31] R.C. Burshtein, V.S. Markin, A.G. Pshenichnikov, V.A. Chismadgev, Y.G. Chirkov, The relationship between structure and electrochemical properties of porous gas electrodes, Electrochim. Acta 9 (1964) 773–787. https://doi.org/10.1016/0013-4686(64)80064-5.

[32] J. Giner, C. Hunter, The Mechanism of Operation of the Teflon-Bonded Gas Diffusion Electrode: A Mathematical Model, J. Electrochem. Soc. 116 (1969) 1124. https://doi.org/10.1149/1.2412232.

[33] M.B. Cutlip, An approximate model for mass transfer with reaction in porous gas diffusion electrodes, Electrochim. Acta 20 (1975) 767–773. https://doi.org/10.1016/0013-4686(75)85013-4.

[34] X.-L. Wang, S. Koda, Scale-up and Modeling of Oxygen Diffusion Electrodes for Chlorine-Alkali Electrolysis I. Analysis of Hydrostatic Force Balance and Its Effect on Electrode Performance, Denki Kagaku 65 (1997) 1002–1013. https://doi.org/10.5796/kogyobutsurikagaku.65.1002.

[35] X.-L. Wang, S. Koda, Scale-up and Modeling of Oxygen Diffusion Electrodes for Chlorine-Alkali Electrolysis II. Effects of the Structural Parameters on the Electrode Performance Based on the Thin-Film and Flooded-Agglomerate Model, Denki Kagaku 65 (1997) 1014–1025. https://doi.org/10.5796/kogyobutsurikagaku.65.1014.

[36] S. Pinnow, N. Chavan, T. Turek, Thin-film flooded agglomerate model for silver-based oxygen depolarized cathodes, J. Appl. Electrochem. 41 (2011) 1053–1064. https://doi.org/10.1007/s10800-011-0311-2.

[37] A. Botz, J. Clausmeyer, D. Öhl, T. Tarnev, D. Franzen, T. Turek, W. Schuhmann, Local Activities of Hydroxide and Water Determine the Operation of Silver-Based Oxygen Depolarized Cathodes, Angew. Chem. Int. Ed. 57 (2018) 12285–12289. https://doi.org/10.1002/anie.201807798.

[38] D. Franzen, M.C. Paulisch, B. Ellendorff, I. Manke, T. Turek, Spatially resolved model of oxygen reduction reaction in silver-based porous gas-diffusion electrodes based on operando measurements, Electrochim. Acta 375 (2021) 137976. https://doi.org/10.1016/j.electacta.2021.137976.

[39] I. Moussallem, S. Pinnow, N. Wagner, T. Turek, Development of high-performance silver-based gas-diffusion electrodes for chlor-alkali electrolysis with oxygen depolarized cathodes, Chem. Eng. Process. 52 (2012) 125–131. https://doi.org/10.1016/j.cep.2011.11.003.

[40] T.G. Tranter, P. Boillat, A. Mularczyk, V. Manzi-Orezzoli, P.R. Shearing, D.J.L. Brett, J. Eller, J.T. Gostick, A. Forner-Cuenca, Pore Network Modelling of Capillary Transport and Relative Diffusivity in Gas Diffusion Layers with Patterned Wettability, J. Electrochem. Soc. 167 (2020) 114512. https://doi.org/10.1149/1945-7111/ab9d61.

[41] F. Mugele, J.-C. Baret, Electrowetting: from basics to applications, J. Phys. Condens. Matter 17 (2005) R705-R774. https://doi.org/10.1088/0953-8984/17/28/R01.

[42] J.T. Gostick, M.A. Ioannidis, M.W. Fowler, M.D. Pritzker, Direct measurement of the capillary pressure characteristics of water–air–gas diffusion layer systems for PEM fuel cells, Electrochem. Commun. 10 (2008) 1520–1523. https://doi.org/10.1016/j.elecom.2008.08.008.

[43] T. Haas, R. Krause, R. Weber, M. Demler, G. Schmid, Technical photosynthesis involving CO_2 electrolysis and fermentation, Nat. Catal. 1 (2018) 32–39. https://doi.org/10.1038/s41929-017-0005-1.

[44] S. Dieckhöfer, D. Öhl, J.R.C. Junqueira, T. Quast, T. Turek, W. Schuhmann, Probing the Local Reaction Environment During High Turnover Carbon Dioxide Reduction with Ag-Based Gas Diffusion Electrodes, Chem. Eur. J. 27 (2021) 5906–5912. https://doi.org/10.1002/chem.202100387.

[45] L.-C. Weng, A.T. Bell, A.Z. Weber, Modeling gas-diffusion electrodes for CO_2 reduction, Phys. Chem. Chem. Phys. 20 (2018) 16973–16984. https://doi.org/10.1039/c8cp01319e.

[46] D. Franzen, C. Krause, T. Turek, Experimental and model-based analysis of electrolyte intrusion depth in silver-based gas diffusion electrodes, ChemElectroChem 8 (2021) 2186–2192. https://doi.org/10.1002/celc.202100278.

Curriculum vitae

Personal details

Name	David Franzen
Address	Falkenstraße 16, 44137 Dortmund
e-Mail	franzen.david@t-online.de
Date of birth	06.07.1990
Place of birth	Münster

Professional experience and education

since 09/2021	**Technical project manager** ARCANUM Energy Systems GmbH & Co. KG
08/2017 – 12/2020	**Research assistant** TU Clausthal Institute of Chemical and Electrochemical Process Engineering
10/2014 – 06/2017	**Master of Science** TU Clausthal Process and Chemical Engineering
10/2011 – 04/2015	**Bachelor of Science** TU Clausthal Process and Chemical Engineering
08/2003 – 06/2010	**Abitur** Graf-Stauffenberg Gymnasium Osnabrück

www.ingramcontent.com/pod-product-compliance
Ingram Content Group UK Ltd.
Pitfield, Milton Keynes, MK11 3LW, UK
UKHW022000190726
13853UKWH00004B/1638

9 783736 975149